FUEL

CARY WOLFE, *Series Editor*

(*continued on page 135*)

FUEL

A SPECULATIVE DICTIONARY

Karen Pinkus

posthumanities 39

UNIVERSITY OF MINNESOTA PRESS

MINNEAPOLIS • LONDON

The University of Minnesota Press gratefully acknowledges financial assistance for the publication of this book from the Hull Memorial Publication Fund of Cornell University.

Published by the University of Minnesota Press
111 Third Avenue South, Suite 290
Minneapolis, MN 55401-2520
http://www.upress.umn.edu

Printed in the United States of America on acid-free paper

The University of Minnesota is an equal-opportunity educator and employer.

22 21 20 19 18 17 16   10 9 8 7 6 5 4 3 2 1

Library of Congress Cataloging-in-Publication Data
Names: Pinkus, Karen, author.
Title: Fuel : a speculative dictionary / Karen Pinkus.
Description: Minneapolis : University of Minnesota Press, 2016. | Series: Posthumanities ; 39 | Includes bibliographical references.
Identifiers: LCCN 2016020934 (print) | ISBN 978-0-8166-9997-1 (hc) | ISBN 978-0-8166-9998-8 (pb)
Subjects: LCSH: Fuel—Philosophy—Dictionaries. | Power resources—Philosophy—Dictionaries. | Power (Philosophy) in literature—Dictionaries.
Classification: LCC TP316 (print) | DDC 662.6—dc23
LC record available at https://lccn.loc.gov/2016020934

CONTENTS

ACKNOWLEDGMENTS

Years ago my application to the University of Southern California Provost's Future Fuels and Energy Initiative was met with a generous grant—much to my surprise. Without this spark I would surely never have embarked on this project, even though it is much changed since that first proposal. The USC College deans also provided support, including an award for a fantastic undergraduate research assistant, JoJo Marshall (now at The Nature Conservancy). Many colleagues, including Jennifer Wolch, Michael Quick, and David Bottjer, encouraged me in what were early days for humanists studying climate change. At Cornell University, I am especially indebted to Tim Murray and Tim Campbell, among many others. A fellowship from the National Humanities Centre in Canberra, Australia, offered me time and space to make leaps ahead. I am thankful to the director, Debjany Ganguly, and for the stimulating conversations with my fellow fellows. Audiences at Rice University, Rensselaer Polytechnic Institute, New York University, the Istituto Svizzero of Rome, Ohio University, and elsewhere gave me much needed feedback. For kind invitations I am most grateful to Peter Friedl, Marina Peterson, Kristin Ross, and James Wilcox. Caroline Levander and Dominic Boyer at Rice facilitated a semester-long visit at the Center for Energy and Environmental Research in the Human Sciences. It was the perfect setting to put the finishing touches on this book as I explored future research. Matthew Schneider-Mayerson, Derek Woods, and many other friends made it a truly magical time. Beth Ahner, Vincent Bruyère, Gökçe Günel, Adeline Putras-Jones, Allan Stoekl, Imre Szeman, Daniel Tiffany, and Maria Whiteman provided advice and support at different moments. Cary Wolfe and Doug Armato of the University of Minnesota Press—you are amazing!

And as always . . . Richard Block, Roberto Diaz, Peggy Kamuf, Annamaria Moiso, Pani Norindr, Renzo Ovan, Sabrina Ovan, Daniel Pinkus, Deborah Pinkus, Tommaso Pomilio, John David Rhodes, Hilary Schor, Joy Sleeman, and Bernard Yenelouis. I write with the fondest memories of Miguel Ángel Balsa and Maurizio Giuffredi. Bob Kaufman has put up with my incessant, anxious chatter about climate change . . . for eons and with no end in sight.

FUEL

A SPECULATIVE DICTIONARY

AIR

A dream: It is September 19, 1783. You are inside the palace of Versailles along with the court (including Louis XIV and Marie Antoinette). Perhaps you establish your location there by glancing at yourself in a mirror in the great hall—as a mise-en-abyme. Or is this too much of a cliché? Perhaps you are dressed in period costume. Or maybe an anachronistic Emma Peel–style aerodynamic flight suit designed for maximum mobility. Or both, in the way that dreams allow. You leave the palace and move into the garden just at the moment that the Montgolfier brothers have finished all of the processes associated with heating air. You breathe in: there is no residual smoke odor. The Montgolfiers are just lifting off, accompanied by several farm animals. Along with the court—oblivious to the preparatory energy-intensive stages— you marvel as the balloon floats upward and then comes back down safely (no animals are harmed in this experiment). Or maybe you are in the basket. Or simultaneously watching from below and floating upward.

Time to wake up.

Air = Nothing

The dream of generating energy from nothing is nothing new. What if we could power our world with free, clean, unlimited, unmetered air? But I have already used two terms—*energy* and *power*—that imply systems. Energy is the fundamental ability to do work. Power is the rate at which energy is used. Fuels, as I hope to distinguish them from systems of energy, are potentialities—a vexing term that I will revisit at various points—perhaps flowing or trapped in rock, perhaps gaseous and invisible, slippery or noxious, not yet rigidified forms of power. Perhaps already discovered, monetized, projected as future earnings; or offset by taxes or compensatory actions, externalized by companies, discounted into climate models *even while still in the ground*.[1]

What follows is a dictionary of fuels, some familiar and in common use, some imagined, some plausible, some the stuff of (science) fiction. As with any dictionary, the reader is free to read any entry, or read them in any order, or read the thing through. Or leave it on the shelf without ever cracking it open.

You won't find "nuclear" in this dictionary. "Nuclear" is not a fuel, and I do my best to retain this rigor, if for no other reason than to provoke thought or surprise or recognition or even fear in readers. To undo a kind of passivity with regard to the place or placement of fuels into vast and interconnected machines, grids, pipelines, storage containers, ecosystems, and even extra-planetary or off-worlds. What does it mean to consider fuel as prior to, as punctual toward, energy and its many infrastructures?

One response to this question is to begin to unravel the knot of "future fuels." Today the oil companies have transformed themselves into "energy companies" in search and development of fuels for the future.[2] Future fuels may be transitional (natural gas), fossil-based, "renewable," fusionable/fission-able, or fantasized substances to come. The American Petroleum Institute has a current project called "Energy Tomorrow," advertising with familiar sorts of images and rhetorics of hope not in the least limited to fuel. Lest we think that this genre is somehow linked to a recent consciousness of end times or catastrophic change, recall a popular slogan from the 1940s: Our times are primitive. True progress is yet to come. Brought to us by the Ford Motor Company. A young Ford watches a waterwheel in a stream in the backyard of a sun-drenched farm as he contemplates the flow of metals on the assembly line in future factories.[3] Or: "There's a Ford in your future." Gaze into a crystal ball and dream of a time when "you'll take your ease in style." You drive your future wife, a rosy-cheeked imp, and an alert terrier in the back seat. Or are "you" the wife who will convince her future husband to spring for the leather seats, maybe to compensate for one of his indiscretions? Or is that man in the driver's seat your chauffeur, taking you on a road trip while your husband's away at the office? A few years of working and saving and all this—fill in the scenario as you like—will be yours.

This future shouldn't be so far away that we forget it, yet we need time to get our act together. Just a little longer. Don't bother me with positive feedback, turning points, thresholds. . . . The very companies that the Petroleum

Institute represents might even develop carbon capture and storage (CCS) divisions that could someday out-earn the extraction and combustion of carbon. Regardless, *these companies* have every reason to blur the distinctions between fuels and energy. The more mystification the better. But genuine "future fuels" never actually come to be, for their time is never any precise moment of political-technological cooperation.

"Potentiality" is a term that will appear at various points in this dictionary. Perhaps the most important thinker to interrogate it recently is Giorgio Agamben, who takes it from Aristotle (*dynamis*) and reads it through the lens of Heidegger, among others. To be sure, Agamben is not interested in fuels or energy in the modern geopolitical sense but, rather, in law and the forms of life it lets be or works to inhibit. In Italian, *potentialità* (along with inoperativity, *inoperosità*, another term that appears in Agamben's writing), feels strange, awkward, as it does in English. Where does "potentiality" stand in relation to "potential" (as in "potential sources of fuel," for instance) or in relation to power? The Romance languages have sets of two different terms for power, both of which are contained in but perhaps also exceed "potentiality": *potenza* and *potere, puissance* and *pouvoir, potencia* and *poder*.[4] English "power" is a powerful term that does not seem to allow for nonpower or the possibility, but not the actualization, of power or the paradoxical nonuse (of fuels). In any case, matters are complex and borders fluid. Because fuels have not yet been inserted into a system that will consume them, use them up, they may = hope. The term "hope" will also reappear throughout this dictionary, but for now let's be clear that it has nothing whatsoever to do with conservation. However much efforts to use "less" fuel (and then, less than what? compared to a baseline calculated during the good ol' days?) on a global scale might produce "fewer" carbon emissions (but then, fewer than what?), conservation is, alas, instrumental rhetoric of the most confounding sort. We might choose to embrace the kind of austerity projected forward in one dark version of a "postsustainable future," to use Allan Stoekl's term.[5] Or we might think with Bataille toward a future when fossil fuels have been exhausted, but forms of human (excessive) energy power the species in new directions. Ten years after Stoekl published *Bataille's Peak*, we are facing another alternate future: one in which we "never" run out of oil. The new North American boom in unconventional oil and gas is changing the geopolitical landscape as *I* write. The immediate present and near-term future colors this dictionary in ways that I could not have predicted when I first conceived of the project. I am not accustomed to writing in "real time" about current events, let alone about events unfolding in decadal timescales (or less), but in fact that is what we are up against. No doubt this dictionary will need to be updated, or at least certain of its entries, but it can at least serve as a reference book that looks both forward and backward from one intensive moment in human history.

Thinking fuel is absolutely crucial to our survival. As a species, of course. But even as "the humanities," now qualified in many institutions with "digital," "medical," or more pertinently "environmental," and even "energy."[6] Writing and thinking about climate change, we feel a strong push from those other experts—the engineers and policy scholars; from students considering their future "fields"; from the scientists who express genuine curiosity about what the humanities can contribute to solutions, or, more likely, to adaptation ("resilience") since this implies a necessary "cultural dimension." Yet— in my own institutional experiences—when the scientists and (especially) the social scientists invite us to the table, they do not want to hear how critical theory might take us beyond, might transgress the limits of what could best be defined as a certain type of value-enriched behaviorism. We may help think about a future, but it is one governed by a tyranny of the practical. Certainly, the pleasure of our company is *not* requested in Bataillian dress. We should probably not speak, between sips of wine, of "unusable energy" that "by definition does not work, that is insubordinate, that plays *now* rather than contributing to some effort that may mean something at some later date and that is devoted to some transcendent goal or principle." We can expect strange looks and awkward pauses if we bring up "the energy of the universe, the energy of stars and 'celestial bodies' that do no work" or the energy "that traverses our bodies, that moves them in useless and time-consuming ways, that leads to nothing beyond death or pointless erotic expenditure, that defies quantification in measure: elapsed moments, dollars per hour, indulgences saved up for quicker entry into heaven." Talking of sacrifice is not only a downer, it will lead to our friends from those more quantitative disciplines looking at their watches and skipping dessert. And then if by sacrifice we mean something like "the movement of the opening out, the 'communication,' of self and community with death: the void of the universe, the dead God" (Stoekl, xvi), the evening will be over rather abruptly. Our survival depends on our ability to be nimble and offer, precisely, something others might call "hope." "No doom and gloom, please," say research centers, funding organizations, and potential collaborators. That will get us nowhere. But where do we think we are going? And powered by what means?[7]

That "future fuels" are perpetually deferred only strengthens the links between hydrocarbons and the present economy. In our present carbon economy, we live under a prevailing view that environmental ethics in the broadest sense (green conscience) and power are antithetical terms that may be forced to come together because of external circumstances. And more mundane, ubiquitous marketers who try to reconcile power and "nature" in the most cynical sense actually only end up strengthening the distinction. Consider, for instance, a recent Volkswagen brochure for "clean diesel"– powered vehicles with the slogan "Go Green without Going Slow." A car speeds along a highway flanked by green hills dotted with wind turbines: "So will you forfeit power in exchange for doing good?" The answer, of course, is no. Yet "no" here, meant to distinguish Volkswagen from its competitors, suggests that power and ethics are, indeed, in the normativity of everyday consumption, incompatible.[8] Or were. Because of course, this all turned out to be a giant hoax. In part, Volkswagen blamed its bad behavior on the stringent requirements of Kyoto and the California Air Resource Board (Lo-CARB—another carbon diet fad?)

A fantasy (courtesy of the Ford Motor Company, 1961): You are seated in the Delta-shaped Ford Gyron. As you take off, the wheels, two in front, one in the back, lift up. A gyroscope helps stabilize the vehicle. You have two passengers. Who? Wife and son? When the daughter arrives, it's time to get a family car. But for now . . . You can visualize your estimated . . . no . . . your *exact* time of arrival and speed on a screen in front of the passenger compartment. Though it is foggy outside, you can clearly survey the road conditions ahead thanks to the infrared rays that project onto your "snooperscope." But that's not all: you can communicate (with the office? or your wife can call her mother?) using a "cordless" telephone!

What's that you say? Fuel? Did I mention the gyroscope for stabilization? Fuel? "The shape and silhouette . . . suggest the possible use of new power sources, such as fuel cells, since it is unlikely that any existing internal-combustion engine both small enough to fit within the front end and powerful enough to propel the vehicle is presently available." Enjoy the glide![9]

Today, Apple is said to be building an electric car (with built-in iPod/iPad/iPhone/iWatch dock, to be sure). And today, "futuristic" vehicles such as the Tesla Roadster bring together money, power, and speed, but again, only through a magical spatiotemporal dislodgment of fuel beyond the frame of perception. Tesla owners, who may feel good about themselves as green consumers, also suffer from "range anxiety," a pathology that the company is working to calm through the blue pill of smart technologies, global positioning software, and interfaces with charging stations. Elon Musk, the company's founder, has announced the arrival of self-driving versions of the Tesla. The self-driving car, center stage at Google's research center, could potentially run on any available fuel. Just imagine the work that auto-piloting will generate for law firms and courts, dealing with insurance, responsibility and indemnity of the humans in or not in the vehicle. Or autonomous vehicles might radically transform the entire economy.[10] All of this talk about smart cars . . . but fuel remains, well, buried. So to speak.

All of this talk about cars . . . when they really contribute only a small proportion of fuel usage, not to mention other sectors contributing to the buildup of greenhouse gases in the atmosphere, such as deforestation, that are not directly related. When "clean," "green" non-carbon-based future fuels will have been "present" (scalable, feasible to the public in the common sense), it will already have been too late with regard to climate change. Notice the awkward use of the future perfect in English in the sentence above! Sober, we witness the contortions of the "future fuel" industry confronting the unfathomable temporality of climate change, of the Anthropocene.

If energy can be understood as a heterogeneous set of self-mystifying systems or machines that block access to thought even as they fascinate us, perhaps fuels are more primal elements that the nonexpert, the nonengineer can attempt to grasp at various points along the way to consumption/combustion. Perhaps this dictionary can help scramble our thinking about fuel—not in order to demonize energy per se, and not in order to create a new hierarchy in which certain renewables take over from fossil fuels (although the fossil/nonfossil distinction must be emphasized), but instead to open up potential ways of interacting with substances (real and imaginary), by wrenching them out of narrative (violently in some cases), and placing them into the form of an idiosyncratic dictionary so they could eventually be replaced by users into new narratives. This is a primary reason why literature and literary language is so important for *Fuel*. The alphabetical form of this book's structure mimics any number of encyclopedias of energy, but its aims are not parody. Rather, because fuels may evade, undo, or destabilize narrative—whether historical/chronological or even fictive—various other forms might do just as well for this discussion.[11]

A series of literary or para-literary texts ranging from antiquity to the present help to position different fuels in relation to one another. Three key terms reemerge at various points: analogy, alchemy, and, as already mentioned, potentiality.

As to the first, it is imperative to distinguish it from a more general figurative field in science, and allegory in particular, as developed by critics like Bruno Latour, Pierre Bourdieu, and Donna Haraway. Allegory, as Bruce Clarke explains in a book centered on nineteenth-century thermodynamics, "has in fact been the perennial model marker of scientific discourses, the primary narrative and pictorial medium through which scientisms operate as tokens of power" (Clarke, 62). And to be more precise: "The angelic and demonic agents in allegories of energy are winged runners in a cosmos techno-science has redescribed as a global communication matrix" (85). On the other hand, Daniel Tiffany, in his book on the nexus tying together poetry, materialism, toys, and weather, notices a "regime of analogy" in modern physics (Tiffany, 4).

In his classic work of literary criticism, *Allegory: Theory of a Symbolic Mode*, Angus Fletcher develops a genealogy or, better, taxonomy of the allegorical as demonic, that is, a figure that embodies one idea, perhaps to the exclusion of others, perhaps in a mode that would, today, be diagnosed as obsessive-compulsive. For Fletcher, we may find allegory in early Christian dramas, realist novels, and science fiction. Sometimes the demonic image (itself unchanging) appears to the reader framed through a progression, a pilgrimage, a set of images on carts. At other times, the reader/viewer "navigates" through space. Allegory is a complex term with many variations, but in the most general sense it refers to something other. And although the distinction is not worth making in absolute terms, fuel, it seems to me, operates or is best captured through analogy, a figure that does not itself move (or move a reader) through time and space, but remains in place like a bridge or link.

Just as energy and fuels are all too easily used without distinction in our everyday speech, so analogy tends to lose its peculiarities, mixed up with metaphor or allegory, which may have a more radically disruptive power, in various genres of writing.[12] As part of an attempt to have commerce with fuels, I will attempt to maintain a certain rigor with regard to analogy, which is also tied to alchemical purification through transmutation.

This dictionary should not be imagined to trace a parallel history of fuels and literature, nor do the texts cited—literary and not—comprise a canon of fuelish writing.[13] Still, a few key moments are worth noting. First, antiquity, because it is here perhaps that we come closest to the hearth, to one origin of "fuel," as I elaborate more fully below. Epics are driven by the force of wind in the sails of the hero. If wind is, today, considered a fuel, it was also a force to drive the hero away from the hearth (fuel). The hero makes use of wood to build his ships. Certainly, in antiquity "wind" and "wood" do not exist side by side in the same category, so any reading of them as fuels back into ancient narratives is willfully anachronistic and perhaps even perverse. So be it. To read *The Odyssey* of Homer as a work of fuels is, naturally, risky. But with risk may come reward as the fuel industry teaches its investors. The *Argonautika* represents a journey around a symbolic object (the golden fleece) that does propel motion away from home. At the same time, scholarship suggests that the fleece was a material object bearing the traces of an alchemical process, the sifting of gold. As both matter and symbol, the fleece enjoys a status as fuel for the purposes of this dictionary.

Other intensive moments arise in the sixteenth century as wood—used to heat homes and cook food, but also to build ships for the British navy—becomes ever more scarce; in the nineteenth century, when energy is solidifying as a science, when industrial production is causing pollution and anxiety about peak coal; and the early twentieth century, when **radium** opens up a utopian way of thinking that transcends consumption and production as they are simultaneously modeled in the mechanized factory. Finally, some fuels are mined from "post climate change" dystopias, descriptions of worlds when fossil fuels exist for minority elites and alternatives circulate in corrupt black markets, spaces radically altered by catastrophic geological shifts. To be clear, though, these are only pressure points, and the through line of this dictionary is more linguistic than historical.

Of all the literary authors that inform this dictionary, none is more important than Jules Verne. There are some historical reasons for this choice, primarily related to his coincidence with the rise of industrialization. On first glance and in the most superficial sense, Verne espouses a bourgeois, Republican ideology. This is one of the first things most critics say about him, although his own biography suggests that later in life he may have experienced ambivalence about science and scientific progress specifically. His works—including *The Mysterious Island* of 1874 (written after coal but before oil)[14]—might be interpreted to reflect some equivocation. He or his works may have contained secrets, as the surrealists suggested (Chesneaux, 20). Still, inasmuch as he is an author associated with a particular techno-philic trajectory, he represents a point of view coincident with the rise of the modern energy state, one, incidentally, in which (free or slave) labor is effaced or neglected in favor of machines that work (or fail to do so).

Indeed, the present dictionary form was partly inspired by Verne's sixty-two *voyages extraordinaires dans les monde connus et inconnus* [Extraordinary journeys into known and unknown worlds] as they tend toward a kind of inclusiveness that compels us to consider all possible forms of fuel, no matter how minor. Verne is extremely fond of lists. In a rather Rabelaisian gesture, a group of men decide to toast to all previous explorers of the African continent in *Cinq semaines en ballon*, the first novel that Verne published with editor Jules Hetzel, who would be an enormous influence on the form and content of his work. The toast—to no less than seventy men (some real, some invented)—becomes a mere series of words, "completely and deliberately out of proportion to the event being narrated" (Unwin, 201). Any order might have been possible, says the narrator, but the alphabetical—chosen at random—is acknowledged to be "très anglais." And while *Fuel* aspires to be more than a list of nouns with the potential to devolve into either nonsense utterances or "a very English" catalogue, I must confess that the absurdity of omitting nothing, inspired by Verne, holds a kind of fascination. And while lists (in any order) may signal mastery, there are moments in Verne when they actually upend the sense of a stable and knowable universe: "By pushing these [lists] a little beyond what seem reasonable limits, Verne frequently manages to destroy that impression of reality that he has also worked so hard to create" (202).

The Mysterious Island reads almost like a dictionary of forms of fuel/energy (Verne does not often make any distinction), but one that is played out in random micro-narratives rather than in alphabetical form. Roland Barthes noted Verne's tendency to render the world "complete like an egg;

he repeats precisely the gesture of the eighteenth-century Encyclopaedist or the Dutch painter" (Barthes, 65). Such prose may seem incompatible with narrative, to be sure. Georges Perec, commenting on a list of fish that goes on for several pages in *Twenty-Thousand Leagues under the Sea*, has the impression of reading a poem (Unwin, 17).

Verne interrupts the rather preposterous narrative of competition over the best flying machine (balloon vs. propeller plane) in *Robur the Conquerer* (1886) in order to instruct his readers that Icarus himself had attempted a birdlike machine (not to mention that the fuel for this machine is **air** or **wind** or a combination of both). He proceeds with a preteritive phrase: "But without going back to mythological times, without dwelling on Archytas of Tarentum, we find, in the works of Dante of Perugia, of Leonardo da Vinci and Guidotti, the idea of the machines made to move through the air" (Verne 1911, 38). As if this list were not enough to establish the centrality of the dream of flying, he continues on for several pages:

Two centuries and a half afterwards inventors began to multiply. In 1742 the Marquis de Bacqueville designed a system of wings, tried it over the Seine, and fell and broke his arm. In 1768 Paucton conceived the idea of an apparatus with two screws, suspensive and propulsive. In 1781 Meerwein, the architect of the Prince of Baden, built an orthopteric machine, and protested against the tendency of the aerostats which had just been invented. In 1784 Launoy and Bienvenu had maneuvered a helicopter worked by springs. In 1808 there were the attempts at flight by the Austrian Jacques Degen. In 1810 came the pamphlet by Denian of Nantes, in which the principles of "heavier than air" are laid down. From 1811 to 1840 came the inventions and researches of Derblinger, Vigual, Sarti, Dubochet, and Cagniard de Latour. In 1842 we have the Englishman Henson, with his system of inclined planes and screws worked by steam. In 1845 came Cossus and his ascensional screws. In 1847 came Camille Vert and his helicopter made of birds' wings. In 1852 came Letur with his system of guidable parachutes, whose trial cost him his life; and in the same year came Michel Loup with his plan of gliding through the air on four revolving wings. In 1853 came Béléguic and his aeroplane with the traction screws, Vaussin-Chardannes with his guidable kite, and George Cauley with his flying machines driven by gas. From 1854 to 1863 appeared Joseph Pline with several patents for aerial systems. Bréant, Carlingford, Le Bris, Du Temple, Bright, whose ascensional screws were left-handed; Smythies, Panafieu, Crosnier, &c. At length, in 1863, thanks to the efforts of Nadar, a society of "heavier than air" was founded in Paris. There the inventors could experiment with the machines, of which many were patented. Ponton d'Amécourt and his steam helicopter, La Landelle and his system of combining screws with inclined planes and parachutes, Louvrié and his aeroscape, Esterno and his mechanical bird, Groof and his apparatus with wings worked by levers. The impetus was given, inventors invented, calculators calculated all that could render aerial locomotion practicable. Bourcart, Le Bris, Kaufmann, Smyth, Stringfellow, Prigent, Danjard, Pomés and De la Pauze, Moy, Pénaud, Jobert, Haureau de Villeneuve, Achenbach, Garapon, Duchesne, Danduran, Pariesel, Dieuaide, Melkiseff, Forlanini, Bearey, Tatin, Dandrieux, Edison, some with wings or screws, others with inclined planes, imagined, created, constructed, perfected, their flying machines, ready to do their work, once there came to be applied to thereby some inventor a motor of adequate power and excessive lightness. (38–39)

Yes, he acknowledges that the list may be a tad bit long, but the reason for it is the laying bare of the various steps in the ladder. For, "Without these attempts, these experiments of his predecessors, how could the inquirer have conceived so perfect an apparatus?" (39). In a sense, the present dictionary takes a cue from what has been critiqued as distinctly unliterary prose.

Verne is also central to the 1966 classic of Althusserian Marxism, Pierre Macherey's *Theory of Literary Production*. In this work, which he later repudiated to a degree, Macherey notes that the central theme of all of Jules Verne's writing is summarized by the motto inscribed on Captain Nemo's submarine, the Nautilus: *mobilis in mobili*, "the future is hidden in the present."[15] I think it is not too much of a stretch to extend this to the core of a notion of "future fuels." *Mobilis in mobili* expresses why, for Macherey, narrative fiction rather than science writing is the ideal medium to discuss the conquest of nature (and for the present purposes, the exploitation of fuels in systems of energy). Reading Macherey reminds us that the future is not some other time and space where renewable fuels will be feasible, for instance, but a discourse that lasts "forever." *Fuel* aims to deconstruct the myth of "future fuels" through its reading of forms, materials, and texts. *Mobilis in mobili*, then, is not just any theme among others for me, but a mode of reading that most powerfully serves to think about fuels in and out of narrative.

So, what is (a) fuel? The word derives from Old English *feuel*, from French *fouaille* or *feuaile*, which in turn derives from the old French, *foaile*, used in the early fourteenth century to refer to a bundle of fire**wood**. Beyond that, *foaile* comes from the Latin legal term *focalia*, meaning the right to demand material for making fire. This right, what Marx will call a "customary law" of the poor, will be severely challenged in the period of transition to capitalism. *Focalia* comes from the neuter plural of the Latin *focalis*, pertaining to the hearth, from focus, or hearth. Recall that Greek and then Latin and the Romance languages had other words for the matter of fire, primarily *pyros* and *ignis*.

Fuel, we might say, begins very early, as a form of combustion inextricable from the hearth, one of the most primeval human traces made on the face of the earth and remaining as archeological ruin. Hunter-gatherers built hearths that may have served to delimit a domestic space, to keep animals at bay (and, incidentally or collaterally, they may also have contributed to illness through smoke inhalation). The hearth predates the torch, by means of which early hominids carried fire outside the home and eventually developed controlled burns as a way of herding prey. If we want to be precise, the earliest forms of anthropogenic "climate change" could be identified with the first hearths.

In the constellation of the Greek Hestia, the Latin Vesta (not etymologically related, despite the apparent acoustic similarities), and the focus and focalia, we find fuel. Hestia is an ambiguous goddess, rarely depicted, and then, perhaps, as a simple, veiled woman holding a staff. Sometimes her attribute is a lump of **charcoal**. Hestia's name may be linked to the Proto-Indo European root *wes*, to dwell or stay. It goes without saying that this is a crucial root for Heidegger in his writings on technology, dwelling, and "the environment." *Wesen* = essence; not in the sense of *quidditas* or a precisely "correct" definition of something although he will pass through this in the case of "technology" in order to arrive closer to something like an "enduring as presence" (*das Währen als Gegenwart*). If this latter sense of essence (rather than a more instrumental definition that attempts to master the object under consideration) is also the essence of Hestia (rather than "what she is" or "what she represents"), then we are, it seems, closer to one—but not the only or absolute—sense of fuel in antiquity.

In ancient epic, the hearth is related to protection, the *oikos* (which is also the origin of both "economy," and much later, "ecology"), home, family, or center.[16] Derrida reminds us that economy is circular: "*Oikonomia* would always follow the path of Ulysses. The latter returns to the side of his loved ones or to himself, he goes away only in view of repatriating himself, in order to return to the home from which [*à partir duquel*] the signal for departure is given and the part assigned, the side chosen [*le partis pris*], the lot divided, destiny commanded (*moira*)" (Derrida 1992, 7). Economy, the epic, could then be thought as destined movements away from fuel, and back to fuel.

The space of the *oikos*—the space, let us say, of fuels before energy—should be understood as opposed to the public and political. The *Oikeion*, Lyotard notes, is a form of writing that is not a form of knowledge and has no public function (unlike economy). "Afterward, yes, when the work is written, you can put this work into an existing function, for example, a cultural function. Works are doomed to that, but while we are writing, we have no idea about the function, if we are serious" (Lyotard 1993, 100). This seems a promising model for fuels that gain power and knowledge only when they are transformed into, inserted into, systems of energy.

When sacrifices are made upon the hearth, they tend to be small, offered to the household gods. In contrast, when the epic hero leaves behind the hearth (however much he may long for it), he sets up altars for large and demonstrative sacrifices during the course of his journey.

Calypso has a hearth inside a cave, where she sings and weaves on a loom with a "golden shuttle" (*Odyssey*, V.62). The cave is located in a grove of alder and poplar. When Hermes visits Calypso, on Zeus's orders, she gives Odysseus an ax and leads him to the "the far end of the island where there are trees, tall grown, alder and black poplar and fir that towered to the heavens, but all gone dry long ago and dead, so they would float lightly" (V.237–40). From these trees—Calypso would have no need of them for fuel since there is no scarcity on her island—he fashions the raft that will take him on his journey home, on his return, *nostos*, to the *oikos*, the hearth kept by Penelope.

In the *Argonautika*, Circe also keeps a hearth. She does not want Medea to linger there:

> Poor wretch, an evil and shameful return hast thou planned. Not for long, I ween, wilt thou escape the heavy wrath of Aeetes; but soon will he go even to the dwellings of Hellas to avenge the blood of his son, for intolerable are the deeds thou hast done. But since thou art my suppliant and my kinswoman, no

further ill shall I devise against thee at thy coming; but begone from my halls, companioning the stranger, whosoever he be, this unknown one that thou hast taken in thy father's despite; and kneel not to me at my hearth, for never will I approve thy counsels and thy shameful flight. (lines 739–48)

Fuel, the hearth, dwelling (as essence, being, building), the circle, the return, the center, *Omphalos*, the economy, the home: all share points in common.

Sometime around the 1300s, "fuel" expands outward from the hearth as place to the materials—the small bundles of firewood—that are combusted. Or rather, the terms are conflated. Fuel begins, then, in a preindustrial world, where homes are heated by wood gathered in forests (and later by charcoal, in London, for instance). Yet atavistic moods around fuel and the hearth persist alongside the beginning of industrialization, as, for instance, in George Eliot's *Silas Marner*.[17] Eliot published her novel in 1861 but set it sixty-odd years in the past. Villagers believe the dwarfish Marner, who spends his days weaving by hand at his own hearth, possesses ancient secrets about the use of healing herbs. Protected in his own space, with his own form of labor, he appears unaware that his is a transitional period. Silas Marner represents, then, an excess of fuel. His hearth "lit up every corner of the cottage—the bed, the loom, the three chairs, and the table" (56). Even after his fortunes increase, he refuses to get a modern oven, preferring the open brick hearth. How does Silas come to be a weaver? His ancestors might have been agricultural workers. But as early as the sixteenth century, English landowners begin to convert arable land—itself created by clearing forests—to sheep pastures, and this leads to a new class of textile workers. The Industrial Revolution depends, in part, on the fact that Britain began to export textiles and, due to the land conversion, to import grain. At the end of the novel Marner returns to his ancestral home, Lantern Yard. It does not register with him that the people who mill around the square once dominated by the chapel are actually factory workers on their lunch hour. He does not understand the immense shift in modes of production that have transformed the world beyond his immediate domestic sphere.

If fuels are related to the home, the hearth, then these words in a minor text by Lyotard may help us locate literature in relation to fuel: "Today, economics belongs to *Öffentlichkeit*, what we call *Wirtschaft* [the economy] is precisely part of the public sphere. . . . If 'economic' means *öffentlich*, it implies that the *oikos* itself has slipped away elsewhere. . . . for me, 'ecology' means the discourse of the secluded, of the thing that has not become public, that has not become communicational, that has not become systemic, and that can never become any of these things. This presupposes that there is a relation of language with the logos, which is not centered on optimal performance and which is not obsessed by it, but which is preoccupied, in the full sense of 'pre-occupied' with listening to and seeking for what is secluded, *oikeion.*"

This discourse is called "'literature,' 'art' or 'writing' in general" (Lyotard 1993, 105). For Lyotard the *oikos* is not a place of peaceful retreat, but rather a space of (familial) tragedy. Lyotard suggests that the place of fuel may not (only) be perfectly calm, but precisely in its turmoil it might be a space of literary writing. Inasmuch as fuel, in the present context, functions like a figure, we can approach fuel in literature, but not with the expectation of mastery through its thematization. In fact, it is precisely for the opposite reason that fuel and literature seem linked—in their undecidability.

Air = (A Figure of) Nothing

In the dream of the free-floating balloon, we have moved far away from the dirty, difficult labor of the extractive industries, so influential as to have shaped collective actions the world over.[18] We are far in time and space from the labor of the automobile industries (so influential as to have given a name to an entire paradigm, not just of physical exertion involved in the assemblage of cars, but of wages and unions, surveillance and living conditions, ideology, even cities, now bankrupt or rebranded), and indeed labor *tout court.*

To call upon **air** as a fuel is to move outside the sphere of the human in the carbon present, to a realm of utopian fantasy; of green fuels and green jobs (whatever else these might mean, they are apparently performed by a multicultural and yet patriotic labor force that finds its sense of community not in a struggle against capital but in its dedication to Nature).

So, what if we could capture **air** as a *fuel*? Air is clean, absolutely renewable, and free. (When we talk of regulation of "air rights," we really mean something more like urban vertical space rather than a physical-chemical element that could be harnessed, like horses, to produce motion.) Air fueled various devices, including some of the machines of antiquity—automata, moving statues of Hero of Alexandria, and later, town clocks, braking systems for trains. Air lifts balloons—flying machines lighter than air—but only when heated. In the late eighteenth century, the Montgolfier brothers believed they had captured a new element ("Montgolfier gas") inside a paper bag filled with hot air. Later, their compatriots replaced air (which cooled very quickly, forcing the balloon to descend) with **hydrogen** gas (which does not exist ready to be utilized as such, but must be created through an energetic process). Moreover, I have said nothing of the fuels employed in the brothers' paper factory needed to produce the bag or *balon*. So air cannot legitimately be termed a "first fuel" of balloon travel. And then once airborne, balloons are driven by **wind**, something distinct from air.[19]

Verne's *Mysterious Island* begins as a group of Americans (Abolitionists: an engineer, a sailor, a journalist, his ward, a boy; a freed slave and a dog—the latter two both loyal to their master) jump into the basket of the fully inflated balloon moored in a public square in Richmond, Virginia. They do not even require combustion in order to move—all they do is loosen the ropes that are keeping the balloon from floating, and the rest is done by the exceptionally strong winds of a ... superstorm ...

WE INTERRUPT THIS DICTIONARY WITH A SEVERE
WEATHER REPORT FOR MARCH 23, 1865:

Surely no one will have forgotten the terrible northeasterly gale that was unleashed at the vernal equinox of that year. The barometer fell to 710 millimeters, and the storm went on unabated from the eighteenth to the twenty-sixth of March. Great was the devastation it wrought, in America, Europe, and Asia alike. ... Shattered cities, uprooted forests, shorelines ravaged by crashing mountains of water, ships slammed against the shore ... In the ranks of natural disasters, it outstripped even the horrific devastation witnessed at Havana and on the island of Guadeloupe, on October 25th, 1810, and July 26th, 1825, respectively. (Verne 2001, 3–4)

Today the French engineer Guy Nègre has developed air cars for city driving. The Airpod is a bulbous two-seater straight from a sci-fi art director's notebook, while it simultaneously refers back to bubble cars of the last century.[20] It could be manufactured from glass fiber and polyester resin or even Styrofoam, for maximum efficiency. The Airpod, futuristic and mod in design, is nevertheless meant for autonomous and immediate (if limited)

FIGURE 1. "Il se recontrèrent près de la nacelle" (The men meet near the vessel), *The Mysterious Island.* Courtesy of the Columbia University Rare Book and Manuscript Library.

mobility. So it represents, for our purposes, a future as a continuation of the present.

Behind the futuristic design, though, as with clocks, brakes, or guns, the air to drive a vehicle must first be compressed, by human force, steam or wood, or by electricity. Air can, in the here and now, perform as a fuel, then, but only after secondary intervention, after the expenditure of another fuel. Behind the clean, green shell of the compressed air vehicle (or the **hydrogen** or electric car) lurk secrets.[21] The fact is that in all of these cases the emission source has simply been displaced, from the tail pipe—where plumes of smoke in graphic design have long signaled speed and power—perhaps to a central electrical plant, just as in a **perpetual motion** machine the first motion is displaced in time so that the machine appears to work by itself. Courtly **automata** performed their marvelous rituals after having been wound up off stage. After a time, they lose their vitality and stare, moribund, at the prince (unless they are retired in time).

If by fuel we wish to indicate a "first matter" or "prime mover," then air can only be fuel—now, in the carbon present—in fantasy.[22]

Rather than bearing a strong significance on its own, the word "air" was modified by the term "ambient" in various early writings. Leo Spitzer, in a book-length essay, recalls that "air" was understood as that which surrounds or goes around (us). It was, for some thinkers, not so much "climate" as the physical or perceptive space where atoms reside (Spitzer, 181). Air may be related to embrace, although in Latin this is less rich than in Greek. For vital materialists, perhaps air really is some thing. It has thickness, it vibrates, it encases or embraces the human as it also has properties of life.

Certainly, words have transformed inert matter into vital matter. They have transmuted sterile metals into fertile wombs that radiate energy, as Blaise de Vigenère acknowledges in his 1600 treatise on fire and salt, part philosophical prose on matter, part recipe book, part instruction manual for physico-chemical experiments. De Vigenère admits to the moribund quality of metals. To be sure, for many early modern writers there is little or no distinction to be made between "metal" and elements that have come to be used as fuels. They may be classed as dead matter, "but poets on the other hand have used them for various sorts of metaphors and figures" (De Vigenère, 123). Does figuration share properties with fuel? If so, then we might be allowed the following analogy? Energy is like a discourse (langue) that is disturbed by punctuations of figurative fuels. Energy, then, might be thought of as a scientific structure that works toward absolute communicability. Scientific discourse (which I am equating with energy) has as its aim to eliminate all that qualifies as figure from its vocabulary and syntax. Unfortunately (from its driving perspective), energy is forced to rely on fuels to help it run, to jump-start it. But someday, energy dreams, I might be done once and for all with figures. And then . . . my power will truly be unlimited. (It will be noted that in this discussion I will often rely on figures—simile, analogy, occasionally metaphor, even allegory—to think about a possible relation between fuels/figures and energy/discourse.

ALBATROSS

Pencroft, the sailor marooned on Verne's mysterious island, is obsessed with the idea of building a ship (to be fueled by **wind**). Even before he begins the process of cutting down timber, he knows that he will only be able to make a small craft, capable of—at most—reaching a small charted island nearby. (That figure 2 presents the book's readers with what appears to be an extremely seaworthy vessel—adorned by an American flag that has inexplicably found its way to the island—is only one of the many contradictions we find if we study the work critically rather than simply letting it flow over us like the ocean waves.) Pencroft's companions fear the risk of any voyage is too dangerous, but a sailor must sail just as an engineer must make machines and an object must move if force is applied to it.

So Pencroft labors. Meanwhile, journalist Gideon Spilett captures an albatross. The boy, Harbert, wishes to domesticate it (to keep it from mobility), but Spilett wants to use the bird to convey a message to civilization:

> Gideon Spilett thus wrote up a succinct message and sealed it inside a sack of heavy gummed canvas with a request that it be forwarded at once to the offices of the New York Herald. This small sack was attached not to the albatross's leg but on its neck for these birds are in the habit of sitting on the water's surface for their rest; then the swift courier of the air was given its freedom, and it was not without a certain emotion that they watched the bird disappear into the fog toward the west.
>
> "Where is he headed?" asked Pencroft.
>
> "Toward New Zealand," Harbert answered. (Verne 2001, 324)

FIGURE 2. The "Bonaventure," from *The Mysterious Island,* built by the stranded "colonists" with their own hands. Are we to believe that this vessel could make a short hop to a nearby island, but not far enough to reach a continental landmass? Once seaborne, what would stop it from continuing on? Courtesy of the Columbia University Rare Book and Manuscript Library.

The colonists will also send a message in a bottle, fueled by the currents of the **ocean.**

In what sense, then, can we justify the albatross as worthy of an entry in this dictionary? Can any object—living or dead—that moves another object be considered a fuel in the broadest sense? The question of self-movement could take us down a rabbit hole to Aristotle and the definition of **soul.** The *De anima* begins with a survey of past thinkers, all of whom, the author asserts, consider both (self-)movement and sensation as key points in defining the essence of the soul. So, for instance, Democritus thinks soul is "a sort of fire or hot substance" like "motes in the air which we see in shafts of light coming through windows" and, like Leucippus, the soul is thus "identical to what produces movement in animals" (I.ii). "Thales, too, to judge from what is recorded about him, seems to have held soul to be a motive force, since he said that the **magnet** has a soul in it because it moves the iron."

This dictionary and its author are far from being equipped to analyze Aristotle in depth. To be sure, in his account of what soul is and who or what might be said to possess it, the philosopher never uses the words "fuel" or even "wood," only "fire" (that is, a substance that moves but also requires a spark to ignite it and combustible matter to continue to move it). This reinforces the idea that while wood—and especially the faggots that were bundled or gathered to make fires around the hearth in antiquity—is a crucial element of matter, it is only once we come to a threshold of its very scarcity, or the idea thereof, that a broader category of "fuel" comes about, perhaps in early modernity.

We might also pose the question of the albatross as fuel to those thinkers oriented toward objects, toward a flattened ontology of a multiplicity of relations between objects, people, animals, and so on. The albatross could be said to be an actor or an actant in a drama of which he is unaware. An actant, as Jane Bennett describes it, following Bruno Latour, can be nonhuman. It must have the ability to "make a difference, produce effects, alter the course of events" (Bennett, x). One might be tempted to say that as actant the albatross is only a vessel fueled by small fish and krill, but how far down the molecular chain should we travel before we have arrived at the "first" or indivisible fuel? We might find ourselves going back to the **sun,** for instance. Or beyond?

Recently researchers have found that they can harness the beating of bird (or bat or even moth) wings to generate enough electricity to fuel devices used for the scientific study of the creatures in question, replacing solar-powered batteries.[23] The Vernian albatross generates power more directly, without the intervening machine, as a self-powering research entity, until—someday, perhaps—we might harness together an entire flock to provide distributed power on a larger scale.

ALCOHOL

The term refers generally to a chemical family (including **ethanol**, that substance much touted in the public sphere in a recent past, green-washed into our collective hopeful brain) with combustible properties. As it is the first liquid in our dictionary, we might pose a more general question about consistency. A liquid can be stored, sold, and put into an energy system in measured doses, but it will flow rather passively so that one "batch" will mix with another. Light sweet crude oil will have the same effect in an engine—with only slight variations—regardless of exterior conditions. It has no say in whether the car drives on the righthand side of the road or goes at 25 miles an hour, or whether the driver has her foot on the accelerator, or whether the car propels itself forward using sensors while a human sends text messages to another human awaiting the arrival of the vehicle at a shopping mall.

In 1916, long before "global warming" or "climate change" were household terms, but under the threat of high gasoline prices and looming shortages, the U.S. House of Representatives considered a bill to establish a government commission specifically for the purpose of "experimentation" with denatured alcohol made from farm waste. The term "experimentation" was much debated during several congressional sessions. European scientists had already proved that alcohol could function well as a fuel with only minor alterations to existing car and farm-machine engines or other infrastructures. Experts had established the proper methods and additives to distill and denature the elixirs. Alcohol seemed to fulfill one of the crucial tasks of "future fuels": a new substance that could simply be inserted into an existing system without either disturbing or undoing its efficacy. In this context, by "experimentation," as more than one witness testified, what was really meant was that Congress would have to make economic calculations about how to subsidize farmers in the construction of stills and distribution of product; about how to regulate the farmers to be sure they were actually denaturing rather than making moonshine. And perhaps most important of all: how to get around the Standard Oil Company, which held a monopoly over the production of denatured alcohol (although for use as a solvent, not a fuel). Or, as another witness who had spent time in Germany, noted:

> The Rockefeller interests have been busy strangling the interest in denatured alcohol, not in this country, but in Germany, France, England, and Canada; but they were not as successful there as they were here. They know that promulgation and introduction of alcohol sounds their death knell for it will enter into every phase of rural and urban life, unifying the two and bring about improvements not dreamt of by the present generation.

This is not the last time we will hear of a fuel so clean (even the temperance unions favor it), so safe (unlike gasoline, it can be extinguished with water; this is important since it might be used in home stoves, and "No matter what kind of a fire or where it is, a woman naturally runs for water and throws it on the fire" [H.R. 43]), and so unlimited (its source is "as inexhaustible as the sunlight and the air" [49]) that it can be counted on to alter collective life for the better in many ways.

Giving farmers the means and regulations to produce fuel on a small scale, for local distribution and consumption, met with significant resistance. It goes without saying that tensions between patriotism ("Drill, baby, drill"), regulation, resource peaks, land use, support of agriculture, and feedstock/crops persist, stubbornly, into the carbon-intense present.

During American Prohibition, enterprising men wrote to Henry Ford to propose the use of potable alcohol as fuel for automobiles. Ford responded in one case: "With the world's supply of petroleum and therefore, of gasoline, fast playing out and the day of alcohol fuel for automobiles and tractors just dawning, the present brewery properties are assured a future much more useful to the community and more profitable to themselves, than has been the case in the past (*Dearborn Star*, November 23, 19??).[24]

On his experimental farm in Dearborn, Ford was actively producing alcohol. Let us recall that Ford felt that consumption of alcohol by workers was to be discouraged actively at every turn. His Sociology Department intervened in the realm of temperance, forming no small part of the larger mode that will be known as Fordism. While Ford ostensibly supported farm life (he constantly stressed his rural origins), and while he certainly attempted to establish vast swathes of land for his use, in Muscle Shoals and in the Amazon in Brazil most notably, he despised farm animals almost as much as "New York bankers."

That alcohol might be extracted wholly from plants that would otherwise have served as feedstock was, for him, unproblematic. If it came from *parts* of plants that were unsuitable for feedstock, thus potentially making farms more efficient, this was a matter of indifference.

From prison ("Americanism and Fordism," notebook 22), Antonio Gramsci asserts that Prohibition produces the new type of man that will also produce a new type of product and new factory labor. The nonuse of a substance (therefore freeing the substance—in theory—to act as a fuel) is crucial to the sociochemical makeup of the work itself.

In Upton Sinclair's *Oil!*, during Prohibition alcohol fuels parties of a new class of men made rich by oil and Hollywood. People talk of cases of Scotch smuggled from Mexico. An "emperor of the screen world" has his basement storage bunker penetrated by clever thieves. It seems the police might have been in with them, so Hollywood man Koski brings in a private detective and threatens a scandal. "By this means he had got back the greater part of his casks and bottles; but alas, the real stuff was gone, they had all been emptied and refilled with synthetic" (Sinclair 2007, 330). Alcohol loosens tongues, and tales are told of an oilman who wasn't making it so he took out an insurance policy and then shot off his own toe while rabbit hunting. Or another who got rich leasing oil land from Indians. But, as a reveler points out, the Indian Chief whose lands yielded oil now has "a different colored automobile for each day of the week, and he figures to get drunk three times everyday" (333). Booze and **oil**, two liquids that fill the tank . . .

A version of the Ford Fiesta from the late 1970s modified to run on 100 percent corn liquor was nicknamed the "boozemobile."

Alcohol might qualify as a fuel in the strictest sense if we thought of bottles sitting on the shelf, doing nothing but waiting either to be consumed or for their contents to evaporate. But if we think of producing (and denaturing) alcohol for the express purpose of inserting it into modified internal-combustion engines or other machines, we can no longer be so certain about its status. Do fuels have a right to be preserved? Do they have or deserve to have ethical responses? Here is Ian Bogost, writing about the rights of what he calls "units" (and note that his example happens to hinge on the operation of fuel in a machine!):

> When I turn the ignition of my car, the engine intake valve draws a mixture of air and gasoline into the cylinder. The piston rises, compressing the mix. Once it reaches the top of its stroke, the spark pug ignites the fuel, detonating the flammable aliphatic compounds within it. The explosion drives down the piston, which in turn rotates the driveshaft. The cylinder's exhaust port opens, and the fume of exploded fuel exits toward the tailpipe. Are these gestures repugnant or reprehensible? Or are they merely thermodynamic, devoid of greater consequence? (Bogost, 74)

And he continues: "When we talk about the ethics of internal-combustion engines, we usually discuss only the first and last steps, the social and cultural practices that encourage driving in the first place, or the plume of combustion gases that exit the vehicle and enter the environment" (75). In other words, in focusing on that particular machine that is the ICE, the fuel itself is left out of the discussion.

When a substance lies in wait, unused, yet usable, as fuel, we might call such a state of waiting "potentiality." This term, as many thinkers have noted, implies or contains power and yet simultaneously describes a state prior to the use of that power. The word *energeia* was first used by Aristotle, according to most scholars. It is built on the word *argon*, meaning act or deed. So *energeia* = enactment. For Aristotle, all matter has *energeia*, which maintains it in being and is the activity of tending toward its natural function or end. The sheer potential for action—before a substance is on route toward its *telos*, is *dynamis*. Clearly, Aristotle's *energeia* is very far from our current, common definition of energy, particularly as we have distinguished it from fuel. This word did not come to mean something like a motive force to drive machines or a form of "power" until much later in Western history.[25]

Here it may be worthwhile to pause to briefly consider Giorgio Agamben's discussion (via Heidegger) of the Aristotelian *dynamis* [or potentiality] and *energeia* [or actuality]. Again, to be clear, the term *energeia* should not be confused with energy in the modern sense. Rather, Agamben returns to Aristotle, suggesting that these two terms concern Being. They refer to a human capacity to "to be able (to) (do something)" or to have a faculty. What does it mean to have a faculty? To "be able" (to)? To have a faculty is to be able to pass into action, but also to not do so. A great deal rides on how we read a complex sentence in Aristotle (in English: "A thing is capable of doing something if there is nothing impossible in its having the actuality of that of which it is said to have the potentiality.") Some have interpreted this as "what is possible is that with respect to which nothing is impossible." Read in this sense, Aristotle seems to form a tautology or a truism. *Dynamis*, potentiality, does not mean a state or even that is not impossible, but rather, is closer to a "capacity." Whether we have a right to anthropomorphize fuels to the degree that they might possess capacity in this sense remains inconclusive. What I find especially compelling here is the notion that fuel or potentiality or *dynamis* is not necessarily inferior to actuality or *energeia* in some hierarchical sense. As Kevin Attell suggests, the former cluster might be anterior and superordinate, richer than actuality (Attell 2004, 92).[26]

Recent work in vitalism and object-oriented philosophy might conceive of certain fuels as forms of life or at least grant them agency. What would this change? Would we humans have more empathy for fuels? Would we cherish (conserve) them more if we thought of them as coexistent actors? For now, as I read Agamben, what distinguishes a human from a fuel or other living matter is that only humans "use" fuels, only humans can make other matter impotential, that is, they can add a dimension of *dynamis* to other beings when these beings are otherwise dumbstruck as *energeia*.

ALGAE (OR SEAWEED)

John Evelyn was a gentleman landowner, a member of the Royal Society, and author of the *Fumifugium*, a 1661 treatise railing against (coal-based) air pollution in London. His 1664 work on **wood**, *Sylva, or a Discourse of Forest-trees, and the Propagation of Timber in his Majesty's Dominions*, was reprinted numerous times and remained popular even to the nineteenth century. The *Sylva* developed out of the milieu of the Royal Society (its motto was *nullius in verba* and some of its members believed that writing was a waste of time). It was dismissed by some as too practical to constitute significant science and yet too rhetorically stylized to lead to actual agricultural reform. In any case, clearly *Sylva* was addressed not to "rustics (meer Foresters or Wood-men)" but to gentlemen (thought to be free from self-interest), and ultimately meant to curry favor with the king ("Advertisement," 1:xcix). The English forests—

"this publique Fruit"—are promoted throughout the work, in part bolster-ing the naval superiority, the right of the Royal Navy to the matter that might otherwise be "fuel." In fact, a main target of the author's criticism are the uncivil "foresters and borderers" who steal wood (2:148).

Evelyn uses the specific term "fuel" to refer to wood gathered in England's forests. Yet at the same time—for perhaps one of the first times in printed literature—fuel can refer to other combustible means. Evelyn remarks on seaweed (*quercus marina*—its very name "sea oak" bears fuel-as-wood), as is found packed in oyster barrels: "This sort of fuel is much made use of in Malta and the islands thereabout, especially to burn in their ovens, and the peasant who first brought it into custom, I find highly commended by an author as a great benefactor to his country" (2:108). Benefactor? Yes, because the peasant has found an alterative to wood (precious for the British navy) and to New Castle coal in a London "poyson'd with smoak and soot"; to free it from "all this hellish and pernicious fog, by furnishing it with a fuel suffi-cient to feed and maintain all its hearths and fires with sweet and wholsom billet" (109).

After providing the reader with information on precisely how the seaweed can be dried and cured, burned on its own or as an additive to "sea-coal fuel," Evelyn continues: "These few particulars I have but mention'd to animate improvements, and ingenious attempts of detecting more cheap and useful processes, for ways of charing-coals, peat, and the like fuliginous materials: as the accomplish'd Mr. Boyl has intimated to us in the fifth of those his pre-cious *Essays* concerning the usefulness of natural philosophy . . . to which I refer the curious" (ibid.).

From Evelyn's premodern Britain to Paolo Bacigalupi's post–climate-change Bangkok . . .

In Bacigalupi's dystopia, *The Windup Girl* (2009), salt water threatens to contaminate algae baths that are growing (secretly) in a factory. Actually, the factory produces what are the primary energy systems available—kink-spring mechanisms wound up with joules and used to power various other forms of production, a factory fueled by calories fed into enormous and nightmarish **Megadonts**. So, secret (alchemical?) algae for what? To be con-verted into biofuel and used in more conventional (that is, from the per-spective of the future, historical) motor mechanisms? Bacigalupi does not convey that information directly. Rather, it seems as if in the novel algae is less a future fuel and more a future *catalyst* to make caloric fuel more efficient. A scientist explains: "Properly cured, the algae provides exponential improvements in torque absorption. Forget its calorie potential [e.g., as foodstuff]. Focus on the industrial applications. I can deliver the entire energy storage market to you, if you'll just give me a little more time" (Bacigalupi, 8). Consequently, a Chinese assistant is tempted to open a safe and sell the secrets:

> The blueprints are there. Just inches away. He has seen them laid out. The DNA samples of the genehacked algae, their genome maps on solid state data cubes. The specifications for growing and processing the resulting skim into lubricants and powder. The necessary tempering requirements for the kink-spring filament to accept the new coatings. A next generation of energy storage sits within his grasp. And with it, a hope of resurrection for himself and his clan. (32)

Flashback: Today, algae is promoted as a "future fuel." "Future," because in the present, energy input is greater than output, but, some experts say, with improvements in the five-step process (cultivation, harvesting/dewatering, lipid [oil] extraction, lipid conversion to biofuel, and the creation of co-products such as methane or animal feed—a process that cannot but strike us as analogous to the processes of alchemy, whatever forms they may take), algae might someday serve in the place of gasoline. It may be grown using little or no fresh water, and it takes up much less space than common biofuels, so say optimists who research this area.

Recently, the founders of a company called Solazyme began growing algae for the development of an oil derivative. This oil has an application "today" as a very expensive topical anti-aging elixir sold under the brand name Algenist (algae + alchemist + geneticist?). The company is also marketing proteins to replace eggs and butter in brioches (Almagine: algae + margarine + imagine?—And how can I get a job making up these names?), until it is able to manufacture large-scale renewable fuels.

In Kim Stanley Robinson's *Forty Signs of Rain*, a scientist is giving a lunchtime presentation in a biotech firm. Her project, she explains in her own words (that is, placed in quotation marks), is to alter the DNA of algae in the hope that it might be made to grow on trees, enhancing the tree's ability to absorb lignin and therefore to act as an enhanced-carbon sink. Other scientists ask questions that prompt the speaker, Eleanor, to respond in clear prose (the kind that the average reader of such a novel might understand): "Many trees host these lichen, and the lichen regulate lignin production in a way that might be bumped, so the tree would quite quickly capture carbon that would remain sequestered for as long as the tree lived. . . . The Lichen's photosynthesis is accomplished by the algae in it" (Robinson, 197). Keeping this kind of prose in mind—that is, a prose that is scientific/didactic, addressed to the nonexpert reader, a time-out from the flow of the narrative, a prose that comprises various portions of the present dictionary—we can return to Verne. For Pierre Macherey, recall, Verne's slight departures from the ur-narrative—Robinson Crusoe or the Island—represent the best way to capture "the future in the present" or "mobilis in mobili."

In a rather uncharacteristic passage from *Twenty Thousand Leagues under the Sea*, Captain Nemo reveals the secret of the Nautilus to his captives (possibly because he believes they will never escape to tell others): "There is a powerful agent, obedient, rapid, facile, which can be put to use and reigns supreme on board my ship. It does everything. It illuminates our ship, it warms us, it is the soul of our mechanical apparatus. This agent is—electricity" (Verne 1993, 77–78). And he goes on to expose the (sole, or one of several?) source of the agent (for this, see **salt**). But Professor Arronax doesn't seem to care (about the fuel, let us say). Instead, he is focused on and skeptical about electricity as a machinic agent: "How could electricity produce such power? What source of energy was he really tapping? Was it the exorbitant voltage developed by a new kind of induction coil? Had he worked out a new transmission system, a secret system of levers that could step up the power infinitely"? (78).

In *The Mysterious Island*, Verne revives Captain Nemo, lost at the end of *Twenty Thousand Leagues under the Sea*. Now revealed or retooled as an Indian prince, having escaped the cruel power of British Imperialism on earth's surface, Nemo has apparently managed to harvest electricity from seaweed to power an underwater crypt/art museum encrusted with carbuncles.[27] To be sure, Verne mentions seaweed only in passing. Yet given his noted comprehensiveness, the fact that neither Nemo, nor the brilliant engineer, Cyrus Smith, leader of the castaways, nor even Verne himself delve into the energetic properties of seaweed suggests that it exists as a possible fuel, but is not terribly exciting. It is not electric.

(. . . Engineer Cyrus Smith, like so many of Verne's heroes, is deemed worthy "not only by the amount of scientific knowledge he possesses, but also by the level of his *énergie*" [Evans, 56]. If energy is, in Verne, both thermodynamic power and masculine virtue, can we extend the analogy to suggest that with regard to energy, fuel is a form of feminine ambivalence?) "The dog, Top, represents the equivalent of electricity in The Mysterious Island. He is a force that helps keep the men energized" [Macherey, 213].)

(On the other hand, energy can also manifest itself as [feminine] excess, almost a form of hysteria. In Arthur Conan Doyle's "The Sign of Four," Sherlock Holmes's "theory" of the crime is contrasted with the supercilious "energy" of the Scotland Yard policeman on the case. Holmes tells Watson: "We have had an immense display of energy since you left. He [Jones] has arrested not only friend Thaddeus but the gatekeeper, the housekeeper, and the Indian servant" [Conan Doyle, 117]. Jones's energy is useful in that he will inevitably leak to the press and thus the real murderer will let down his guard, thinking he is not suspect. Of course, Holmes's "theory" will later prove that all of this expenditure of energy was completely in vain.)

. . . To return, then, to Verne's narrative: "Cyrus Smith instructed his companions in all things, and he especially explained to them the practical applications of science. The colonists [that is his term for the men, not colonialists, but those who found a colony, a campsite, a confraternity] had no library at their disposition; but the engineer himself was like a book— already ready, always open to the exact page they needed, a book that solved all their problems for them and that they leafed through regularly." Generally, in Verne's novels, "scientific knowledge is always encyclopedic in nature— learned through accumulation and then transmitted through regurgitation" (Evans, 41). Strangely, as with the aforementioned seaweed, in the more general case of electricity, Verne often remains rather vague. In fact, in *Twenty Thousand Leagues under the Sea*, the professor asks Nemo, precisely, for an explanation of electricity. Nemo's initial response almost sounds as if he thought electricity was a fuel rather than a system of energy: "There is one source of power that is obedient, rapid, easy to handle, and flexible enough in its application to reign supreme on board my ship. It does everything. It gives me light as well as heat, and it is the very soul of all my machinery. This power-source is electricity." He then goes on to clarify that he is able to achieve great power by the use of "Bunzen cells." Electricity passes through large electromagnets that drive a propeller shaft. However, the professor is left with a series of questions, not least of which is the source of the power, that is, the **fuel** that generates electricity. For Verne, electricity was a future (fuel) force, not only a technology to come, but "a short-cut towards the mastery of nature" (Chesneaux, 43). But above all, in its vagueness, electricity "also avoided the necessity of having to provide an answer to awkward questions about the role of labour in a modern mechanized world" (ibid.).

Similarly, in *Robur*, the inscrutable visitor to a Philadelphia men's club disdains the **air** balloon without actually revealing what fuel is used for *his* machine: "Robur had not availed himself of the vapor of water or other liquids, nor compressed air and other mechanical motion. He employed electricity, that agent which one day will be the soul of the industrial world. But he required no electro-motor to produce it. All he trusted to was piles and accumulators" (Verne 1911, 42). We should not expect Jules Verne—who was not interested in or able to describe large-scale industrial machines—to share our interest in distinguishing fuels from energy or even to express absolute rigor in his catalog of technologies. As one critic notes, "for the average reader, it is enough to know that the magic of electricity propels the Nautilus— the remainder is assumed to be part of the uncommon engineering genius of Captain Nemo" (Evans, 71).

Today, electricity is the largest sector of U.S. energy use, yet the most-used fuel—oil—does not contribute to it.

. . . meanwhile, we have lost track of seaweed, almost as if Verne himself forgot that he had toyed with it as a fuel . . .

AMBER

The Greeks called it *electron*, referring to the fact that it attracts other matters to itself when rubbed (static electricity). It was also called ἅρπαξ; and χρυσοφόρον from its golden color. One might well ask if amber deserves a place in a dictionary of fuels. Perhaps not, since static electricity is minuscule in terms of energetic quantity. In the form that the Greeks found amber, it would not have been able to drive a piston or lift a mechanical arm. It is incompatible with even the measurable daily output of a slave. Still, it is a matter that exercises force on another matter and in this sense it might be understood as analogous to **radium**, to give just one example.

"Scale," a key term in recent thought around the Anthropocene, seems crucial here. For a substance to qualify as a fuel, does it have to achieve a minimal level of kinesis or combustion? What would determine this threshold?

Like **coal** or **bitumen (jet)**, amber was washed up in seawater. So natural philosophers wondered if these substances shared amber's electricity. In a treatise on the property of magnetism written in 1600, physician to Queen Elizabeth I, William Gilbert, notes:

> Many modern authors have written about amber and jet as attracting chaff and about other facts unknown to generality, or have copied them from other writers: with the results of their labors booksellers' shops are crammed full. Our generation has produced many volumes about recondite, abstruse, and occult causes and wonders, and in all of them amber and jet are represented as attracting chaff; but never a proof from experiments. . . . The writers deal only in words that involve in thicker darkness subject-matter. . . . Hence such philosophy bears no fruit. (77)

So, in order to determine if coal and jet share the properties of amber, it is not enough to speculate esoterically with words. We might say that in the *De magnete*, *figures* act as fuels in lieu of the substances themselves, which, if subjected to experimentation, would then take the form of *discourse* in which they would reveal themselves as lacking attractive properties. In the early modern scientific treatise, figures and discourse stand in analogous relation to fuels and energy. Gilbert's treatise may be compared with another one from slightly earlier, Agricola's enormous undertaking, the *De re metallica* (1556).[28] It should not surprise us in the least that both authors intersperse figurative language with precise descriptions of recipes or experiments, lists of nouns, and printed images of tools. Gilbert continually interrupts his discussion of amber to criticize the philosophers who cite in Greek and Latin but do not have any proof. Both texts express a tendency toward modern scientific prose that is interrupted or punctuated by poetic or narrative—that is, literary—elements. (Of course, novelist Jules Verne is cut from the same cloth.) Ultimately, as Gilbert notes, in order to determine the attractive property of a given element or substance, we should not rely on language or on the "crowd of philosophers" with their different, predetermined classes of attractions. For such a class cannot by any means content us, nor does it define the causes (*causas*) of amber, jet, diamond, and other like substances, which owe to the same virtue the forces they possess; nor of loadstone or of other magnetic bodies, which possess a force altogether different from that of those other bodies, both in its efficiency and in the sources whence it is derived. We must, therefore, find other causes of movements, or must with these stray about as it were in darkness, never at all reaching our goal. (80)

Gilbert then goes on to elaborate the ways that amber does not attract (e.g., not by heat; not by the dissipation of air, pace Lucretius; not by kinship with an essential quality to other substances). Amber attracts, rather, because it is an "electrick." This can be understood only in relation to a larger study of geology of the period. In brief, Gilbert believed that there were two matters in the earth from which all others drew their properties in different proportions: the fluid and the humid. Like **jet**, amber is compacted moisture:

All bodies are united and, as it were, cemented together by moisture. . . . But the peculiar effluvia of electrics, being the subtilest matter of solute moisture, attract corpuscles. Air, too (the earth's universal effluvium), unites parts that are separated, and the earth, by means of the air, brings back bodies to itself; else bodies would not so eagerly seek the earth from heights. (92)

Some electrics attract even more powerfully when rubbed or polished. Gilbert deduces that they have some peculiar quality (aside from the air) that makes them attract other bodies. He eliminates various possibilities, suggests experiments to his readers, and interjects rejoinders about the crowd of philosophers who have done neither.

Anti-matter—See Dilithium

Automaton

The thirteenth-century French goldsmith Guillaume Boucher is supposed to have created an automaton for the Great Khan, a spectacular fountain composed of gold lions whose mouths spouted wine, milk, and other drinks. An angel perched on the top of a tree played the trumpet, but Boucher was forced to work the mechanism by hiding a man underneath the contraption.

The man would have blown air into the trumpet at the appropriate moment. But a malfunction caused his breath to trigger the liquids to flow, perhaps spraying unwitting courtiers.

A man hidden in a box beneath Wolfgang von Kempelen's eighteenth-century automaton chess player controlled the actual moves.

Visitors to the General Motors Pavilion at the 1933 Chicago World's Fair are amazed by a mechanical Chief Pontiac who "breathes, moves his head, eyes and mouth, describes your costume and answers your questions about the Pontiac Economy Straight Eight" (GM brochure, 1933).

A Japanese "windup girl" performs her marvelous rituals at a sex club in Paolo Bacigalupi's eponymous novel. The label ("New People" is the more politically correct moniker) refers to a previous generation, when sex toys were literally loaded (by human hand) with joules. But as we learn, Emiko is a later-generation development, a genetically engineered humanoid, pre-programmed with instincts to obey and please. Still, her name bears the history of her kind: "She moves toward the stage with the careful steps of a fine courtesan, stylized and deliberate movements, refined over decades to accommodate her genetic heritage, to emphasize her beauty and her difference. But it is wasted on the crowd. All they see are stutter-stop motions. A joke. An alien toy. A windup" (Bacigalupi, 36). So, technically speaking, what fuels her? She exists on her own caloric motility. In this figure, we locate a disjoined genealogy of the courtesan: from a wound-up automaton in the premodern period, to the puppets-without-strings of modern science fiction: the Marias of the Metropolis jolted into action by electricity or the replicant Prins; then a return (in the past—our future) to the premodern, a radical technological-epistemological break due to the catastrophic break-down of biogeochemistry as we know/knew it and the end of fossil fuels . . . and then a new beginning . . . in the area of genetics.

In Michelangelo Antonioni's 1962 film on environmental devastation, *Red Desert*, the protagonist, Giuliana, moves around the port of Ravenna as a kind of automaton, wound up (just barely) offscreen.[29] Her alienation is existential. Petroculture dominates the landscape (industries, architecture, moods), but there are no fuels on screen per se.

Giuliana tosses and turns in bed, agonized by her doubts and fears. She is awakened by a sound—ambient and electronic. At first she (and we) cannot locate its source. We may assume it is part of the soundtrack. But soon the sound becomes diegetic when Giuliana enters her son's bedroom and sees his robot—made from an erector-set kit, probably assembled by her son and his engineer father—moving on its own power. What is uncanny about the scene is precisely the fact that the son is asleep—at least this is what we can verify, for his eyes are closed and he appears immobile. So we have to conclude that the robot started on its own. Perpetual-motion machines, like erector sets that helped turn boys into men in the 1960s, must also be wound up, even if they may go on for a very, very long time. Still, a Vaucanson waits in the wings . . .

In his discussion of man and machine in the first volume of *Capital*, as elsewhere, Marx is not particularly concerned with fuel. It is not a term he uses, in part because it bears no direct significance for human labor power in the factory. For instance, the distinction between a tool and a machine is significant only with regard to motive power. All machines have a motor mechanism, a transmitting mechanism, and a tool-element, morphologically like that which was at one time wielded by a human hand. The motor mechanism either generates its own motive power (Marx gives the examples of a steam, caloric, or electromagnetic engine) or receives its power from an already existing natural force (water or wind, for instance—both have their limitations, as Marx notes, because they are not always readily available or consistent!). The engine is certainly more reliable, for it can be installed in any factory, without the capitalist being forced to rely on proximity to a stream or variable weather conditions. Nevertheless, the Marx of *Capital* does not investigate further the differences between natural sources of power and mechanical modes of factory operation. He measures them with regard to productivity, but without recognizing what appears a staggering disparity, making them ill-suited for comparison. This is precisely because Marx's uncanny elasticity allows him to shift points of view, and at this point in his argument, he thinks as a capitalist. What is important is that once set in motion, the machine performs actions with its tool-elements that were formerly done by workmen:

> In the first place, in the form of machinery, the implements of labour become automatic, things moving and working independent of the workman. They are thenceforth an industrial *perpetuum mobile*, that would go on producing forever, did it not meet with certain natural obstructions in the weak bodies

and the strong wills of its human attendants. The automaton, as capital, and because it is capital, is endowed, in the person of the capitalist, with intelligence and will; it is therefore animated by the longing to reduce to a minimum the resistance offered by that repellent yet elastic natural barrier, man. This resistance is moreover lessened by the apparent lightness of machine work, and by the more pliant and docile character of the women and children employed on it. (272)

The so-called self-acting mule appears to workers in the factory like a **perpetual motion** machine because they do not see the steam engine that powers it, let alone the coal that is burned to produce steam or the mines where coal is dug. "The site of the mystery has been displaced from the human element to the mechanical element; it is as though matter were copulating with itself, and its activities, which are at once 'mathematical' and enigmatic, become something that approximates to the primal scene of the technological imagination" (Doray, 46).

The distinction between tool and machine, then, is one of quantity, but not quality. The fuel pumped into the factory is hidden or elided because the capitalist cares primarily about the mechanics of the production of surplus value and labor power in relation to the modern machine. Then the prime mover in Marx's writing is not "first matter" as fuel, but a machine that regulates the rhythm of production:

"A system of machinery, whether it reposes on the mere cooperation of similar machines, as in weaving, or on a combination of different machines, as in weaving, or a combination of different machines, as in spinning, constitutes itself a huge automaton, whenever it is driven by a self-acting prime mover," Marx writes (1967, 381). The machine may or may not run on fossil-based fuel. It's of no particular importance in the morphology of the tool-machine. What matters is that the machine, which could or should ease the burden of the laborer, ends up expanding labor time for the capitalist. To underscore this point, rather than describing an early fossil-energy infrastructure, Marx invokes preindustrial machines via Aristotle:

"If," dreamed Aristotle, the greatest thinker of antiquity, "if every tool, when summoned, or even of its own accord, could do the work that befits it, just as

the creations of Daedalus moved of themselves, or the tripods of Hephaestos
went of their own accord to their sacred work, if the weavers' shuttles were to
weave of themselves, then there would be no need either of apprentices for the
master workers, or of slaves for the lords. (ibid., 408)

A passage from the *1844 Manuscripts* may help us grasp the importance of
this observation for *Fuel*. In a section titled "The Accumulation of Capitals
and the Competition among the Capitalists," Marx copies various passages
from the political economists regarding fixed capital or machinery. About
Ricardo, Marx comments: for him, "nations are merely production-shops;
man is a machine for consuming and producing; human life is a kind of
capital; economic laws blindly rule the world. For Ricardo men are nothing,
the product everything" (Marx 1959, 18). We are very far from any fear of
that the machine might lead to unemployment and worker unrest. On the
contrary, for the capitalist, an automaton is no existential threat. It is perfectly acceptable as long as it produces at a profitable rate. Then Marx cites
the political economist Jean de Sismondi: "Nothing remains to be desired
but that the King, living quite alone on the island, should by continuously
turning a crank cause automatons to do all the work of England."[30] Without
delving into issues around the relationship of the king to the capitalists (not
to mention the larger organism of the state), we can appreciate the image
Marx conjures. To be sure, this imaginary island is far from that of Robinson
Crusoe (where the single, sovereign, resourceful, and determined survivor
never actually worries about fuels) and far from Verne's Lincoln Island (where
the men make imaginative use of all available fuels while self-governing as a
brotherhood).

BANANA (SEE ALSO **BIOMASS**)

November 29, 1916. Letter to Henry Ford proposing to use bananas grown
on a plantation in Honduras. Reply of December 13 states that Mr. Ford does
not intend to purchase any banana lands "as we have enough waste material
in our country for all alcohol purposes."[31]

You're heading down the jetway at a major airport. As you wait, hoping
with all your might that your neighbor for the flight will not disturb you
with cries, snores, spilling, or annoying chatter, you pass a recent series of

advertisements for HSBC financial. You notice an image of wind turbines made entirely of banana peels. "There will be no difference between **waste** and energy," reads the text. We are to think about the future with and through this financial services company, on our way to catch a plane. In this future— worthy of our investment—bananas serve a symbolic/aesthetic function as the infrastructural elements linked together in an organic-inorganic nexus.

BIOMASS

The term is a broad one, referring to any organic material that can be processed for fuel. **Bananas**, for one. A great deal of biomass today comes from agricultural **waste** products—wheat straw, pine chips, lawn clippings, indigestible husks of sugar cane or corn; or plants that might otherwise be used as food stock (corn), but also plants grown specifically for fuel such as switch grass. Jatropha, a drought-resistant plant, might be the jet fuel of the future. And so on.

In many instances, such matter is converted into a form of **alcohol**, but it deserves its own entry in our dictionary because the forms of fuel are varied. In a strict sense, the materials of biomass are renewable: grown widely, standing in fields, ready to be harvested. Plants are, simply put, solar energy. It is worth recalling that the raw materials of the textile industry were cotton and wool. So before **coal**, the industry most closely associated with the Industrial Revolution was fueled directly or indirectly by the sun. We might end our entry here. Investigating further, however, we are forced to admit that before "it" (and recall that biomass is a broad term for a heterogeneous set of matters) can become fuel, biomass must undergo processes of transformation; perhaps energy-intensive, perhaps not. Such processes may be crucial for thinking about fuel and hope.

Consider **ethanol**, a fuel that is regulated heavily by the U.S. government in relation to both production and consumption. In fact, during the devastating 2012 Midwestern drought, corn farmers, meat producers, and biofuel refineries began to seriously question the mandate by the government that the percentage of ethanol in gasoline be increased each year. Ethanol has been marketed as a green, clean fuel (until its efficiency is considered compared with gasoline—and until it is burned.[32] Production of ground-level ozone, a greenhouse gas, is a serious problem, for instance, but this dictionary of fuels is not necessarily obliged to think that far into the process).

Cellulosic ethanol is a promising *technology*: the input includes plant biomass, plant waste (corn stover, cereal straws, sugarcane, sawdust, paper pulp), human waste (municipal solid waste), agricultural waste, agricultural products grown for energy (switch grass, hybrid poplar trees), and even industrial waste. Once it comes out on the other end of a multistage (patented) process of transmutation, the matter is broken down into lignin, cellulose, and hemicellulose. In a convertor, it undergoes a process of hydrolysis, making sugars, which are then subjected to yet another stage: microbial fermentation, to produce ethanol (and, alas, CO_2). And as with alchemy in the broadest sense as practiced and written from antiquity through modernity, there may be significant obstacles in the process. Lignin, for instance, inhibits the breakdown of the plant matter. Certain acids or enzymes may be added during the phase of hydrolysis to help the breakdown (akin to the secret matters added during *putrefactio* or *nigredo*, in some premodern alchemical canons).

Biofuel conversion passes through multiple stages of burning, cleansing, and redemption of matter, depending on the particular process. Anaerobic thermal conversion, for instance, may take place in a closed oven and in a nearly closed loop.[33] Sometimes the process can be slow. It is often messy. Precise regulation of temperatures and the order of stages of transmutation are themselves carefully guarded as intellectual property, subject to patents— to which material is subject before it can emerge as fuel/capital/green gold.[34]

The production of fuel from biomass raises, then, the (embarrassing) question of **analogy**; of analogy with **alchemy**, so common in everyday speech as to risk being evacuated of all (energetic) power.

A headline from an article on a Houston-based biomass company: "Scientists say they can spin straw into fuel."[35]

And then we consider the instruments used in the transmutation itself, which, if placed side by side with illustrations of the athanors (ovens) or the alembics (stills) of early modern alchemy, could appear indistinguishable.

Verne once described *The Mysterious Island* as a "chemical novel." Can we take from this only that the men engage in experiments to develop the tools and materials they need to survive? But, then, why not call it a novel of mechanical engineering? Or does the novel also veer into the territory of the alchemical narrative, a narrative of fall and redemption?[36] Perhaps the chemical aspect of the novel refers to Nemo and his secret interventions more than to a particular branch of science. At times it seems that for Verne chemistry meant elective affinities, mystery, and suspense rather than a mode of naming, knowing, and predicting the behavior of elements based on their consistent material makeup.

Cyrus Smith needs explosives and asks his men to become chemists for a day. Of course they comply. They would be happy to be—for him—"dancing-and-deportment-masters" if required (Verne 2001, 163). Smith oversees the piling of branches and pyritic schist like a house of cards. Wood inside is lit and then the whole is covered with earth and grass. Vent holes are formed to allow steam to escape. The process resembles that for making **charcoal**. After twelve days, the vessel is opened. As in so many descriptions of alchemical processes, the matter inside is removed and subjected to a secondary process:

> Once the fire had completely reduced the pile of pyrites, the resulting matter, consisting of iron sulfate, aluminum sulfate, silica, charcoal residue, and ashes, was shoveled into a water-filled basin. They stirred this mixture, let it rest, then poured it off into another vessel, obtaining a light-colored liquid that held the iron sulfate and aluminum sulfate in suspension. . . . Finally, once the liquid had partially evaporated, the crystals of iron sulfate precipitated out, and the mother liquor, that is, the unevaporated liquid, containing the aluminum sulfate, was discarded. (165)[37]

Next, the iron sulfate crystals are incinerated in another covered vessel, perhaps more like an athanor, allowing the sulfuric acid to emerge as a vapor that can be condensed in the liquid form as nitric acid. Finally, Smith takes this substance and mixes in glycerine in a double boiler (the alchemical *bain marie*) to make an oily, yellowish liquid. This last step Smith does alone (for it could be potentially dangerous), but the result—when he presents the **nitroglycerine** to his men—is that of a magician emerging with his final trick, see nothing up my sleeves—voilà!

The *athanor* remains one of the most carefully guarded secrets of the tradition. The Medieval alchemist Nicolas Flamel comes upon a book by "Abraham the Jew" written in strange figures, an instruction manual for alchemy. Flamel is able to understand almost everything, and he even attempts the Great Work numerous times. As he notes, while it is commonly called "oven," there exist many names for the athanor: "It has been called a sive, dung, balneaum Mariae, furnace, sphere, the greene lyon, a prison, a grave, a urinal, a phioll and a bolts-head. I called it a house and habitation of the poulet" (59). The author goes on to note that even if he had the proper ingredients to complete the stages of transmutation, even if he had found the prime matter, even if he had constructed the perfect laboratory, without having come across the book, he would never have achieved transmutation because the fire in the oven must be kept at a very precise temperature. Thus, merely being in possession of the apparatus is not enough. One must also know the secrets of its workings.

Agricola has no problem revealing trade secrets, including different types of ovens for smelting and assaying metals in the *De re metallica*. Agricola was a doctor, like the great alchemist Paracelsus. His writing is contemporary with alchemy, yet its style and patronage differ significantly: The Duchy of Saxony stands behind him as he extols the moral and use value of metals, taken from the earth and exploited for the betterment of human life. Perhaps, we might say, he suffers from no sense of ambivalence toward the matters.

Consider a patent description for a process of cracking, that is, refining crude oil for gasoline: an American chemist and metallurgist, Paul Danckwardt, just before the start of World War I, acknowledges that various

FIGURE 3. From Georg Agricola, *De re metallica*. Illustration of an oven. Courtesy of the Columbia University Rare Book and Manuscript Library.

methods have been tried with various degrees of success, but the biggest obstacle to an efficient distillation is the "difficulty of constructing an apparatus which would allow the reaction to go on undisturbed" with "high pressure and high temperature." The precise temperature is truly important. Without it, "the evil [that is, gas and carbon rather than gasoline] was enhanced and the process furnished not the desired products, and the still suffered so that it had to be replaced in a short time, as otherwise the risk of an explosion became too great."[38]

We are forced to take seriously the alchemical analogy in relation to (modern) fuels, I believe. For if we do not, if we dismiss it as a mere (ab)use of a series of figures from the past that add up to a phrase like "magical transformation of materials," we would find ourselves in the embarrassing position of having to confront an embarrassing term: "magical." Do we intend "magical" as metaphorical in the sense that the production of fuels may astound one who is not familiar with the process? Such a reduction does not seem adequate to comprehend the ubiquity of the alchemical through the fields of fuels, and energy, the discourses of value, first matters, separation, redemption of materials, and spirit. By contrast, I will, throughout this dictionary, wish to avoid any sort of mystification of processes that are not only physical but tied so intimately to capital that they can often be said to shape it. Thinking with alchemy poses this fundamental challenge . . .

Ignoble raw material is subjected to a series of processes that break it down and change its chemical composition until it is usable as (noble) fuel. The material itself poses obstacles to the process. The scientist who triumphs must be patient (and pious? Fortunate? With a pure **soul**?).[39]

Many of the writings on alchemy, particularly from what may be called its golden age—the seventeenth century—are purely explicative of a process (such writings may be termed exoteric as opposed to esoteric). But throughout the genealogy of alchemical writing, we find texts that contain tiny kernels of narrative within them. As we enter into a period of modern chemistry, "alchemy" persists, rather stubbornly, outside of narrative, as a marker for something like "magical transformation." In the context of modern, corporate science, it continues to serve this function. But there is a risk that the invocation of "alchemy" will burst out beyond the boundary of something like a marvelous trade secret, to undermine the serious work. So, we might say that "alchemy," in the context of the alternative fuel industry, is a term of potential high risk (with potential high payoff). For some green venture capitalists, such a high risk may be precisely worth taking. For investors on fixed incomes or those seeking wealth preservation, "alchemy" may not have a place in their portfolios.

We will have occasion to revisit the centrality of alchemy to any thought of fuel throughout this dictionary. We must also consider, more broadly, the relation between (the science of) fuels and (the figure of) the analogy.

Analogy, as I noted, has long been a highly significant tool in the development of the scientific method. Recall Gilbert, struggling to understand the class of electrics (including **amber**). Many philosophers believed that amber and **diamond** are like (analogous to) the objects attracted to them, "like one another, but not the same, near to one another in kind," and this is why they are attracted to each other. "But that is reckless speculation; for all bodies are drawn to all electrics, save bodies aflame or too rarefied, as the air which is the universal effluvium of the globe. Plants draw moisture, and thus our crops thrive and grow; but from this analogy Hippocrates in his book *De Natura Hominis*, I, illogically infers that morbid humor is purged by the specific value of a drug." For the early modern scientist, analogy is like (*simile*) inference based on faulty logic. "Men of acute intelligence, without actual knowledge of facts, and in the absence of experiment, easily slip and err" (Gilbert, 82). So in this context, analogy is to be avoided, since observation and repetition of phenomena is the only correct way to proceed. This explains why—skipping ahead—the avant-garde might feel confident in embracing analogy as the placing together of two apparently entirely different elements. Surrealism is predicated on a certain degree of analogy, for instance.

The Greek word *analogia* was a mathematical term meaning "proportion." In fact, Latin sometimes translated it as *proportio*. It was, in this sense, a relation of measure shared by two things, a term of measurement used in a way that is measured (not "fanciful" as Ernest Rutherford will worry when he thinks about the alchemical analogy in scientific language).

"Analogy" actively performs the work of fuel in an analogy about fuel in the work of Filippo Tommaso Marinetti, leader of the Italian Futurists. Speed is his "new religion-morality." Fuels should be stolen (with a certain violence, ideally—launching of star cannons; assaulting 1830 Groombridge, a star in the constellation in Ursa Major) from the earth and heavens. Marinetti would never accept the (passive) **sun** as a fuel: "Guilty as sin are those cities lost in the past where the sun moves in, makes itself comfortable, and doesn't move another inch."[40] Sunlight tries to stop speed, so man must fight back. In this piece, Marinetti is the anti-Heidegger. He exalts the immense tubes that rip (*strappare*) electrical power from waterfalls just as he exalts all violence. At night he disdains sleep and prays to his electric light to keep him awake: "The immense speed of a car or plane allows us to embrace and rapidly confront different places from earth, that is, to achieve mechanically the work of analogy. He who travels a lot acquires genius mechanically and brings the distant close. By gazing at them systematically and comparing them he discovers profound sympathies. Speed is the artificial reproduction of the analogic intuition of the artist" (134, translation mine).

Now we might say analogy refers to a cognitive, not a physical, process, where a subject transfers something from a source, that is, the analogue, to a target. The analogue, then, is what is known, stable, before us. It does not travel or fuel travel (as in Marinetti). It is we, through language, who move. But in a more narrow sense, analogy comes to mean a form of logic in which we take a quality of something and apply it to another thing without, however, subtracting it from the source. Finally, as in Gilbert's sense, analogy comes to mean the relation of similitude between the source (what was at one time an analogy) and the target. The problem is that for Gilbert, analogy may be a false logic in science. It is not enough that two things behave in a similar manner, they must be proven similar by experiment or accepted as dissimilar.

BISMUTH

A slightly radioactive element, bismuth has long been considered as the highest-atomic-mass element that is stable. In fact, the half-life of its isotope, bismuth-209, decays more than a billion times more slowly than the age of the universe itself. Today bismuth has no application as a fuel. Indeed it has few applications, except in pharmaceuticals such as Pepto-Bismol or Kaopectate, where far from a fuel, it acts to block transmission of matter through the intestines. Miners in the age of alchemy—an age when the very extraction of materials (fuels, metals) from the earth was understood in the most profound sense to be an act of transformation—also gave bismuth the name *tectum argenti*, or "silver being made," still in the process of being formed within the earth.

In H. G. Wells's work of fictional science, *The World Set Free*, published in 1914, bismuth plays an important role, not as a primary fuel, but a transitional one. For the purposes of his narrative, Bismuth is one of the first matters that radium transforms into—certainly useful, but not as powerful as the later fuel-state of the element that will power ships and weapons capable of destroying the world.[41]

BITUMEN (OR JET)

Ancient marine life, algae and plankton degraded by bacteria. It is both a (fossil) fuel and a substance with various practical applications (asphalt for roads, roofing tar).

The word "mummy" is derived from an Arabic word for bitumen: it was used in sealing the wrappings of the dead. It may have been used to seal the Tower of Babel (Nikoforuk, 16).[42]

In a key text of seventeenth-century alchemy, *The Chemical Wedding of Christian Rosenkreutz*, adepts awaiting induction into the Order of the Golden Fleece are asked to place an egg into an athanor inscribed with the letters OBLI.TO.MI.LI K. I. VOLT. BIT. TO. GOLT. Perhaps this means: "Prescription: Take pulverized and liquefied fusible bitumen; through music and fire the form of the bitumen is elevated to **Gold**."[43] Afterward, the egg hatches into a bird that undergoes further stages of transformation. Finally, it is decapitated and burned. Its ashes will be used to help resuscitate a pair of royal homunculi, completing a cycle that elevates the adepts to a high level of prestige. This is not a fuel in our sense of the term, but it is matter that undergoes a transmutation in order to cause other forms to take shape. It makes an analogy with fuel.

An eighteenth-century botanist, John Macoun, called it "the ooze."

Bitumen is too thick to be sucked or pumped out of the ground like crude oil, so it is either mined or dug out "in situ." Bituminous sands, or oil or tar sands, are loose particles saturated with petroleum.[44] At present, making liquid fuels from oil sands requires energy for steam injection and right now refining and this process generates two to four times the amount of greenhouse gases as conventional oil. And this is to say nothing of the water required for the process or other forms of "collateral damage." The so-called well to wheels emissions of bitumen are 10–45 percent higher than conventional crude oil. Is there any sense in which, when it seeps up through the ground in the La Brea Tar Pits in Los Angeles, or oozes, untouched, in the Canadian wild, bitumen could remotely be thought of as hope?

And what to say about the Keystone XL Pipeline? What is the deep relationship between this proposed (at the time of this writing) infrastructure and the bitumen that awaits extraction? Where does this stand with regard to "potentiality" in all of its complexity?

In the *Argonautika*, Athena weaves a tunic for Jason to help seduce Medea. Various scenes are depicted in the tapestry, which we might call an ancient (digital) text of warps and weaves. Does Medea read the tunic? Does she interpret individual scenes, or does she merely gaze on it as a general visual field? If she is merely dazed by the overarching sense of splendor, the author, Apollonius, feels compelled to extract from the general field a number of specific scenes, including one depicting "the Cyclops seated at their imperishable work, forging a **thunder**bolt for King Zeus; by now it was almost finished in its brightness and still it wanted but one ray, which they were beating out with their iron hammers as it spurted forth a breath of raging flame" (lines 730–34). This scene is one of potentiality. It is highly likely that the Cyclops would complete the thunderbolt, and certainly we would be surprised if they did not. As beings, the Cyclops "can or can not" finish their work. But in the tunic, this scene can never be finished or else we would know nothing of the process. The tunic-scene allows us to visualize potentiality in action in way that would not be possible in narrative.

CAMPHENE

Turpentine mixed with alcohol, it was used for lighting during the nineteenth century and taxed very heavily to help pay for the U.S. Civil War. It has explosive qualities that make it less than desirable as a fuel.

CANDLE—SEE **TALLOW**

CAROLINUM

Bearing certain properties that resemble **plutonium**, named for a fictional King Charles, **carolinum** is the explosive, radioactive fuel used in bombs in H. G. Wells's *The World Set Free*. The bombs, launched from atomic-powered planes, do not explode immediately upon reaching their targets. They may wait for a period of days before they begin to spread radiation. After nuclear war, the world powers collect all sources of the element and keep them for peaceful purposes.

CHARCOAL (SEE ALSO **WOOD**)

Used widely beginning in antiquity (an attribute of the hearth goddess, Hestia), charcoal can be considered a "first fuel" inasmuch as it not an alloy or hybrid. It is wood that has, however, been subjected to a primary process of combustion (usually in a cone-shaped pile or pit) before it will be combusted again in an efficient secondary process. It was used widely in the British Isles for smelting iron ore and refining iron in forges, for malting and other purposes that required a smokeless fuel. For many such purposes, it was superseded by coke or *anthracite*.

While Agricola describes its use for smelting ores in great detail, charcoal in John Evelyn's *Sylva* is couched in rather secretive and atavistic language. It seems out of place with some of the more modern and political discourse of the book. And while Evelyn does not use the term "alchemy," it should be clear that the process described reverberates within a larger discursive field in which matter (metallic, in particular) is believed to ripen in the earth's womb, imbued with certain qualities depending on the respective influence of fire, water, air, and earth. His language is infused with the alchemical, although he is not writing directly in that tradition. Evelyn notes, for example, that the Venetians are said to have a method of burying timber until it develops a very hard and dry coal-like crust: "*ut cum onmis putrificatio incipiat ab humido*" (2:83). Putrefaction is traditionally one of the stages of the alchemical process before the cleansing or purification and redemption of matter/man. For instance, Nicolas Flamel insists for pages that the corruption of matter is crucial for the Great Work to be accomplished. He couches this phase of the process in allegorical terms, evoking a black man and the decapitation of a black crow. To achieve putrefaction is not something open to everyone. In *his* work, John Evelyn may simply mean to cite (without identifying an author) ancient wisdom to support the fact that with their secret method, the Venetians have managed to sufficiently dry their wood in order to avoid decay. Yet the phrase evokes the alchemical, not in a direct figurative link (analogy, simile, metaphor), but in a tonal one. Charcoal is ancient—like fossils—"I my self remembering to have seen charcoals dug out of the ground amongst the ruins of ancient buildings, which have in all probability, lain cover'd with earth above 1500 years" (85). Deep time, early modern style.

In his Upper Michigan Peninsula sawmill, Henry Ford produced charcoal briquettes (still marketed under the brand name Kingsford) from defective logs or wasted timber. These were burned to power steam engines or heat the workers' houses. The current process for making briquettes follows a complex series of washing the material, filtering out impurities, and reconstituting it. The alchemical analogy transfers smoothly to the factory.

Immediately after World War II, and before petroleum imports resumed again, car manufacturers considered charcoal as a fuel for automobile. Perhaps the engines could have doubled as grills for tailgate parties.

CLATHRATES—SEE **METHANE HYDRATE**

COAL (SEA COAL)

Called "sea coal" in early modern England (because it was believed to wash up in the sea; and to help to distinguish it from charcoal), coal is **peat**, formed from vegetal matter compressed underground over millions of years. More specifically, it is primarily the plant's lignin, left behind after the rest of the plant gets eaten away by fungi.[45] Coal can be scratched out from superficial deposits or dug out deeply from below. Actually, a variety of different hydrocarbons comprises the one, black substance we call coal. Carbon dioxide, of course, but also perhaps elements of sulfur dioxide, nitrogen oxides, and mercury compounds.

Coal—burned in one-room huts or teepees by Native Americans, in the homes of the bourgeois and the miners in northern France, in the factories of London, in Dickens's fictional Coketown, even in home furnaces in parts of the world today, to say nothing of industry—is, it goes without saying, dirty. But aside from the blackened interiors and polluted air it created even in early modernity, its carbon intensity—that its combustion releases large quantities of CO_2 trapped over geological eons, a gas that is neither dirty nor visible—this is the major question around coal with regard to climate change. It is also the reason we must remain vigilant about the term "clean coal" as we hear it in everyday speech.

Coal, mined at Coneygre near Dudley in the West Midlands in the early eighteenth century, was burned in an engine to produce steam, to drive a pump that removed water from the mines. The Newcomen, an "engine to raise water by fire," itself ran on coal that was otherwise not useful for sale, and as it underwent improvements it allowed for ever-deeper penetration into less accessible mines. The Newcomen continues alongside the machines—away from the mines—that burn coal to power machines inside the factories.

By 1800, the British coalfields had standardized equipment and methods. The Newcomen engine pumped water out of the ground, miners used safety lamps, wrought-iron chains and sire ropes for haulage, and iron rails underground for transport. The fields provided domestic fuel for locals as it was used, more and more, for iron-making (railroads and other industrial tools are related to the rise of coal). A vast machine of different inputs and outputs built from below the ground up. The colonists in Verne's *Mysterious Island* are fortunate to find coal near the surface of the earth.[46] Given their small number, they could not—they would not—have the resources to develop an extractive industry, or to mimic one so profound as a form of narrative time and a whole new relation between the body and the earth. Instead, the colonists dig out surface coal and cart it to their "home" (apartments they have carved out of a granite rock face). For the men of Lincoln Island, it is just another fuel, like so many others, standing in wait for them.

No doubt this too represents a fantasy. A long digression into the combustible efficiency or the mining of coal would not make for suspense even if Verne's colonists would be capable of constructing a machine for boring, a system of ventilation, timbering and tubbing to keep seams open, warning signals for "firedamp" or pulleys for descent and ascent, and so on.

A different kind of prose does penetrate deep into the subsurface. . . . In the investigation of the intimate relation that pertains between fuels and humans, it is impossible to overestimate the importance of Zola's 1885 novel *Germinal*. The author visited the Anzin mines in northern France during a strike in 1884 and immersed himself in the culture of coal.[47] His phenomenology of the mine is exquisite. It should be noted that this novel is not autonomous—it was part of the planned twenty-volume "Rougon-Macquart series" focused on questions of illness and heredity.[48] Zola believed firmly in the idea of *milieu* or *ambiance* (environment) as inextricable from the human.[49] "One no longer studies man as simple curiosity . . . detached from ambient nature (*nature ambiante*)," he wrote (cited in Spitzer, 216). Thus, while in the novel those who dwell above ground and tend gardens, for instance, may enjoy relative health and economic autarchy, Zola makes clear that the subsoil is not a foreign realm where man is forced to slave, but in a sense is his proper home. Man dwells underground—it is his ethos, his "essence" (*Wesen*) in the Heideggerian sense. Modeled in part on the epic voyage below the surface of the earth (Dante in particular), *Germinal* enjoyed an afterlife beyond its material borders. It is, above all, a novel about the struggle between labor and capital—embodied in the absentee directors, the petty bourgeois managers, and small business owners around the fictional town of Montsou (money mountain). Miners flocked to Zola's funeral and shouted "Gérminal" as a generalized code word of protest.

Zola describes mining in free indirect discourse, through the eyes of the protagonist, Étienne. In this way, he is the focus of the reader's attention, but not fully identified as the empathetic hero, and this has significant implications for the way language, the human, and coal are made inextricable. It is imperative that we arrive on the scene with Étienne, from outside, so that we become acclimated to the world of the mine over the course of the novel just as we cannot fully lose ourselves in his particular struggle to survive.

Quite soon he understands (so we understand) how the mines enter the bodies of the miners, who are also consumed by it. The mine entrance is a mouth; tunnels lead to its insatiable belly. ("The pit gulped down men in mouthfuls of twenty or thirty and so easily that it did not seem to notice them going down" [Zola, 37].) Underground, the colliers tap on the rock face, loosening coal into tubs. After they open a seam, they fill in the voids with timber, moving another natural substance—another fuel—from the surface down into the caverns.[50]

Étienne meets Bonnemort, an old-timer who is now assigned to surface duty. Over the course of his years in the mine, Bonnemort has become a geological being. As he speaks, flaming coals "cast a gleam of blood-red light across his pallid face" (11). He is prone to coughing fits. "Is it Blood?" asks Étienne. Bonnemort replies, "It's coal . . . I've got enough coal inside this carcass of mine to keep me warm for the rest of my days. And it's five whole years since I was last down the mine. Seems I was storing it up without knowing. Ah well, it's a good preservative" (12).

Through the labor process—and let us recall that one of the central griev-ances in the novel is that the miners are paid only for coal taken from under-ground, not for the ancillary but necessary activities such as timbering—bodies are intertwined with the coal to a degree that they cannot be said to exist as separate entities. For instance:

> Each man hacked into the shale bedrock, digging it out with his pick. Then he would make two vertical cuts in the coal, insert an iron wedge into the space above, and prise out a lump. The coal was soft, and the lump would break into pieces which then rolled down over his stomach and legs. Once these pieces had piled up against the boards put there to retain them, the hewers disap-peared from view, immured in their narrow cleft. (39)

"Ghostly shapes" move in the coal, "and chance gleams of light picked out the curve of a hip, or a sinewy arm, or a wild-looking face blackened as though in readiness for a crime." (40)

And similarly, toward the end of the novel, when the revolutionary, Souveraine, goes underground literally to undermine the past work done, he is possessed:

> He attacked the tubbing at random, striking where he could, with the brace, with his saw, suddenly determined to rip it open and bring everything crash-ing down on his head. And he did so with the ferocity of a man plunging a knife into the living flesh of a person he loathed. He would kill it in the end, this foul beast that was Le Voreux, with its ever-gaping maw that had devoured so much human fodder. (463)

Later, the mine and all of the manmade machinic apparatus finally collapses in what we might call a geo-anthropogenic catastrophic event:

> Le Voreux shook slightly, but it was stoutly built and held firm. But a second shock followed at once and a long shout came from the astonished crowd. . . . From then on the earth never ceased to shake, and there was tremor after tremor each time the ground shifted beneath the surface, like the rumblings of an erupting volcano. . . . In less than ten minutes the slate roof and the head-gear fell in, the pit-head and the engine-house were split asunder, and a huge

gap appeared in the wall. Then the noises stopped, the collapse halted, and once again there was a long silence. . . . It was all over: the vile beast squatting in its hollow in the ground, gorged on human flesh, had drawn the last of its long, slow, gasping breaths. Le Voreux had now vanished in its entirety down into the abyss. (480–82)

We do not witness the collapse of the mine through Étienne, who is buried below. Rather, this is an occasion for Zola, author, with a genuine interest in geology, a fascination with new ideas of geological time, to express a more globalizing vision of his Neptunist, catastrophist theory. Le Voreux's end, he explains, is "a reminder of the ancient battles between earth and water when great floods turned the land inside out and buried mountains beneath the plains" (504).[51] We need to be reminded of this because we may have become too immersed in the struggles of the miners, too empathetic.

In other words, in the novel, coal and human/animal flesh meld to create a cyborg, a figure of zōē (the simple fact of living common to all living beings—in Agamben's influential definition) and bíos (a form of way of life proper to an individual or group) combined. One cannot live without the other—they are literally geo-bio-dependent. Coal cannot be used without using it up (it is not a renewable source of energy); the miners cannot live without work, without using up their lives.

The miners are not only producers of coal, they also consume it, albeit in controlled circumstances. "Every month the Company gave each family eight hectolitres of escaillage, a type of hard coal collected off the roadway floors. It was difficult to light, but, having damped down the fire the night before, the girl had only to rake it in the morning and add a few carefully chosen pieces of softer coal. Then she placed a kettle on the grate and crouched in front of the kitchen dresser" (22). The miners' homes are filled with the smell and dirt of coal. They are immersed in it at all times, whereas the bourgeois enjoy a central heating system, and when they do have coal burning, it is contained "cheerfully" (76) behind a grate in a kitchen that smells of freshly baking brioches.

Like his model Zola, Upton Sinclair spent time in the mines of the Rocky Mountains and witnessed labor disputes before composing his 1917 *King Coal.* Like Étienne, Sinclair's protagonist Hal is an outside observer, an intellectual (not one of the "ethnic" types) who comes to immerse himself in the mine. And like Étienne, Hal surveys the surface of the landscape and contemplates geological time before his journey down:

> As one walked through this village, the first impression was of desolation. The mountains towered, barren and lonely, scarred with the wounds of geologic ages. In these canyons the sun set early in the afternoon, the snow came early in the fall; everywhere Nature's hand seemed against man, and man had succumbed to her power. Inside the camps one felt a still more cruel desolation—that of sordidness and animalism. There were a few pitiful attempts at vegetable-gardens, but the cinders and smoke killed everything, and the prevailing colour was of grime. (Sinclair 1994, 21)

Sinclair, through the free indirect speech of Hal, reports of miners as "a separate race of creatures, subterranean, gnomes . . . stunted creatures of the dark."[52] And he continues: "After Hal had squatted for a while and watched them at their tasks, he understood why they walked with head and shoulders bent over and arms hanging down, so that, seeing them coming out of the shaft in the gloaming, one thought of a file of baboons" (ibid., 22). Again, coal fuses with the human. As in Germinal, the mine disaster is "a thing of human flesh and blood" and miners lay on their backs, trying to catch drops of water from the ceiling to keep alive. Compared with Zola, Sinclair is less interested in a subterranean, Bergsonian, and Deleuzian-becoming, which Keith Ansell-Pearson has described in a book aptly titled Germinal Life, and is instead focused on the manifest issues around actual politics.[53] Still, and most significant, the biopolitical life-form that develops in both novels is a collective being, just as mining is a collective form of labor. No individual body can exist as such in the mine.

The links between coal, coal mining, and alchemy seem crucial to thinking through or with this fuel.[54] A long-standing belief held that the first matter of metals could also be used to create the Elixir or stone of alchemy. The Guhr (or gur) theory of the genesis of metals was popular in the sixteenth century. Guhr, a viscous liquid mud composed of metallic sulfides and sulfates, gives off heat, hence it was believed to be in the process of turning into solid metal. Early modern authors explained the incredible diversity of metals and minerals by the celestial influence on a prime matter in the mines. Metals (and coal would appear in this category) were understood as living entities, growing in the womb of earth from different exhalations, and under the influence of the planets and guhr helped to prove this.

Moreover, underground matter was thought to exist in a relation of analogy or affinity with matter on the surface. Wood and coal (subterranean forests) mirror each other. William Gilbert notes that the hidden, primordial elements of metals and stones lie concealed in the earth, as those of herbs and plants do on its outer crust. Here is another authority writing on the question: "Mineral coals, like the other minerals, were bestowed with their special seed by God in *prima creatione* so they will be nourished, multiplied and propagated to the end of the world" (Bunting, cited in Sieferle, 184).

Yet coal distinguished itself in many ways from plant life. Almost as soon as coal was mined regularly in Britain, authorities understood that it was not a renewable source of fuel. Anxiety about the finitude of coal seems nearly simultaneous with coal's development into power. In 1836, to counter discussion of export, theologian/geologist William Buckland wrote:

> We are all fully aware of the impolicy of needless legislative interferences; but a broad line has to be drawn by nature between commodities annually or periodically reproduced by the Soil on its surface, and that subterranean treasure, and sustaining foundation of Industry, which is laid by Nature in strata of mineral Coal, whose amount is limited, and which, when once exhausted, is gone for ever. As the law most justly interferes to prevent the wanton destruction of life and property; it should seem also to be his duty to prevent all needless waste of mineral fuel; since the exhaustion of this fuel would irrecoverably paralyze the industry of millions. (cited in Sieferle, 190)

Several decades later, in *The Mysterious Island*, Jules Verne stages a conversation among his colonists. The world will eventually come to an end, it is suggested, when coal, the most precious mineral, will run out. Verne's engineer, Cyrus Smith, agrees with the premise that coal is indeed to be valued most highly: "Nature herself has proclaimed it so, by creating the **diamond**, which is nothing other than pure crystallized carbon" (Verne 2001, 326). So does this mean that we will burn diamonds? No, for they are too rare. The engineer predicts that ever more efficient machines for drilling, and then, extraction in Australia and America, will yield coal for the world for at least 250 to 300 years (until approximately 2174 ACE, that is), by which time it will have been replaced by another fuel. "Let us hope so," the journalist says, "for without coal, no more machines, and without machines, no more railways, no more steamships, no more factories, no more of anything that the progress of modern life requires" (326).

See **water** for Verne's solution to scarcity of coal . . .

In *The Windup Girl*, (diesel) coal fuels the vehicles of the army. It is a sign of their immense power that they have access to a fossil fuel.

COBALT 60

A radioactive isotope created synthetically in a nuclear reactor. Used experimentally as an automobile fuel during the "Middle East Oil Crises" of the 1970s in a Stirling (*external* combustion) engine. The decay of Cobalt 60 produces heat, thought the future-oriented engineers (no consideration, then, of climate change, of course), which could be converted to electricity to power an onboard generator, but in theory even solar power could be used.[55] No need to change the form of the car or modes of driving or living. Just take this nineteenth-century design and project it forward. So said future-oriented engineers.

COKE—SEE COAL

DEUTERIUM

Of all the radioactive matters under consideration for nuclear fusion in the real world, before first contact with extraterrestrial life or fuel forms, deuterium (with tritium plasma, from lithium) seems most promising.

Some scientists believe it is just a matter of time before fusion moves beyond the break-even point (that is, eventual output of energy will be greater than the input). In the shorter term, fusion/fission hybrid reactors are more realistic. Fusion is a process, of course, not a fuel. It is a dream that haunts the future . . .

Deuterium is almost limitless in the sea.

DIAMOND

A natural sign of **coal**'s preciousness. Can it be considered a fuel? Diamonds are pure carbon dioxide, so they could, theoretically, be burned as a (fossil) fuel except that they are very hard, not to mention that they are extremely rare given the precise geological conditions required for their production. In fact, most diamond becomes graphite or evaporates into the atmosphere as carbon dioxide. In the very long run, diamonds are a greenhouse gas . . .

The Life Gem Company will transmute hair or ashes of a deceased loved one into a diamond that can be worn as commemorative jewelry.[56] The process, analogous with alchemical transmutation, involves combustion, but the diamond is the end product, not the matter of combustion itself. On the finger and set in gold, perhaps it serves as a *memento carboni*.

Diamond is one of the materials being considered in the design of fuel capsules for inertial confinement fusion (ICF) experiments at the National Ignition Facility, Lawrence Livermore National Laboratory. ICF uses high-powered lasers to vaporize a target capsule containing fusion fuel, creating an implosion.

DILITHIUM

NERD ALERT: Legions of *Trekkies* know that Federation vessels make use of "dilithium crystals" to achieve "warp drive."[57] More specifically, if you insist, the dilithium in the starship engines is composed of **deuterium** and lithium-6 (a stable molecule of lithium) in a crystalline structure—hence dilithium crystal. Technically, dilithium is a molecule with two covalently bonded lithium atoms, while lithium-6 features six bonded atoms, but we can forgive the *ST* writers a little poetic license. Technically, dilithium is not a fuel in this dictionary's more rigorous sense of the term. It is a porous medium that moderates matter-antimatter (fusion) reactions, but it deserves an entry here because it is sought in order to make movement of the ships possible. A number of plots early in the *ST* franchise revolve around dilithium mining. Later, it can be replicated as simply as Earl Gray tea. Dilithium was "played" on the small screen by quartz.

The real fuel, if we like, that allows the Enterprise to boldly go, is **antimatter**.

In the present, real world, the production of antimatter requires enormous quantities of energy and space. In the *Star Trek* universe, it is produced efficiently on board by "quantum change reversal devices" that flip the charge of particles (Krauss, 93).

EROS'S ARROW

Although we might commonly say, for instance, that the arrow fuels Medea's love for Jason, we may find here that the use of fuel as a verb is not adequate to "fuel" as a noun. Drawn back in the bow, the arrow of Eros is potential energy:

> And quickly beneath the lintel in the porch he strung his bow and took from the quiver an arrow unshot before, messenger of pain. And with swift feet unmarked he passed the threshold and keenly glanced around; and gliding close by Aeson's son he laid the arrow-notch on the cord in the centre, and drawing wide apart with both hands he shot at Medea; and speechless amazement seized her soul. (*Argonautika*, bk. 3, 275–98)

The poet of the *Argonautika* does not stop here. He must extend a series of figures out in time and space:

> But the god himself flashed back again from the high-roofed hall, laughing loud; and the bolt burnt deep down in the maiden's heart like a flame; and ever she kept darting bright glances straight up at Aeson's son, and within her breast her heart panted fast through anguish, all remembrance left her, and her soul melted with the sweet pain. And as a poor woman heaps dry twigs round a blazing brand—a daughter of toil, whose task is the spinning of wool, that she may kindle a blaze at night beneath her roof, when she has waked very early— and the flame waxing wondrous great from the small brand consumes all the twigs together; so, coiling round her heart, burnt secretly Love the destroyer; and the hue of her soft cheeks went and came, now pale, now red, in her soul's distraction. (ibid.)

Finally, then, while the bow and arrow may be better classified as a system of energy, the effects of the arrow are like fuel on the hearth. In this form, then, the machine works like discourse, along a straight line, whose primary and secondary (this writing here and now) effects work like figures.

Consider, then, another popular (extended) analogy to explain the difference between potential and kinetic energy. Potential energy, it is often said, is like money in the bank. It stands ready to be spent. So potential energy is to a rich man as kinetic energy is to a poor man (Clarke, 44).

(The problem here is that if a rich man had money in the bank and spent it, he is likely to be able to get more money.... Perhaps he has a virtually unlimited source of family money. Perhaps he used his money to buy goods that he can then sell at a higher price. Finally, the analogy only works as long as we consider it very broadly, like a maxim or schoolbook phrase. But once we unpack it, we find that it raises more questions than it answers. Analogy has the potential to undo physical clarity even while it pretends to serve as a pedagogical tool.)

ETHANOL—SEE BIOMASS

FLEECE

Although the fleece is warm and golden—it might even radiate heat—in its various renderings, it does not seem particularly combustible. Still, the fleece is assumed to be a symbol of royal **power**, hence Jason's uncle, Pelias, seeks it to legitimate his seizure of the throne from Jason's father or to help consolidate his sovereignty. In *The Argonautika*, then, the fleece is the object of a quest fueled by the desire of Pelias to destroy Jason, fueled by the oracle who foretold that Pelias would be destroyed by a man with one sandal, fueled by the winds (themselves fueled in part by Hera's desire to destroy Pelias for failing to honor her with sacrifices), themselves fueled, in turn, by the sacrifices made upon hearths and altars to Phoebus, Apollo Embasius (god of embarkation), and others. The Argos is built by the "fuel" of Athena's will.

In Apollodorus, the origins of the fleece lie in female deception (paralleling the deception of the women of Lemnos that is told as part of the Argonauts' journey). That is, in an earlier, indeed a very ancient myth that Greece may have borrowed from oriental sources, Ino deceives her husband, Aeolus, into believing that he must sacrifice his son Phrixus (from another wife, Nephele) in order to overcome a drought of wheat. Nephele gives her son a ram with a fleece of gold. The ram carries away Phrixus and his sister Helle (who falls from the ram, hence the Hellespont). The ram takes Phrixus to the Colchians on the Black Sea, where the ram is sacrificed and the skin is nailed to an oak tree by the Colchian King Aietes. A substitute for a human sacrificial lamb, the ram is sometimes said to be adorned with colored wood or with a necklace.

The fleece is much discussed, but it barely appears in the epic as anything material—only for a brief moment as Jason seizes it from the dragon. Just as quickly, it is forgotten as Jason struggles with Medea. Is *The Argonautika* a drive, a dangerous quest desired by the gods, that circles around an empty object?

Perhaps Apollodorus avoids describing or lingering over the fleece in detail, since it may have been only a bit of animal skin washed in alluvial gold and used to line panning troughs, as was the practice in Colchis. So all of the attention to the fleece in mythography (and then the mystical orders named for it in the Christian West) amounts to a bit of mineral deposit on an animal skin?[58]

A later tradition suggests that the fleece was not actually an animal skin or golden, but was, instead, a piece of parchment containing instructions for the alchemical transmutation of gold. It is not far, then, to a metonymy that understands the very gold of alchemy as the fleece. In his treatise, Nicolas Flamel warns anyone who might be blessed enough to "conquere this rich golden Fleece" to "thinke with himselfe (as I did) not to keep the talent of God digged in the Earth, buying lands and possessions, which are the vanities of this world." Rather, the "fleece" should be donated to charity (Flamel, 36).

Young nobles found worthy after a series of moral and physical trials in the alchemical narrative of the *Chemical Wedding of Christian Rosenkreutz* are initiated into the Order of the Golden Fleece.

In Pier Paolo Pasolini's *Petrolio* (1975), an epic and alchemical Great Work named for, centered to some degree around oil, the (fallen?) hero, Karl or Carlo, travels to the Middle East in search of oil-as-fleece in a series of notes titled "The Argonauts." Taking this into consideration, we are forced to ask: Does the analogy of the quest itself, that is, disembodied from a mystical object, extend to the materiality of oil?

Carlo attends cocktail parties where bored bourgeois wives make small talk. "Jason's plans for capturing the golden fleece are actually mere intentions," he hears (Pasolini, 143). Everyone is talking about Medea's love for Jason. "As for the golden fleece, it shines with an oily, almost palpitating light, very red, here and there in the desert" (144). Note 54 is in fact titled "The real voyage to the Middle East," and it recounts the details of the failed Moroccan investments of one of the subsidiaries of ENI, the Italian state fuel entity. The company was unsuccessful in finding oil in the Sudan and Eritrea, we learn. Ten years later, Carlo returns, this time as head of the commission. He has a "foundation in dreams" and is considered an expert in the Orient. He visits familiar places and moves freely. Here the prose is a mix of Orphic poetics, dreams, exotic-erotic descriptions of the desert, and the bureaucratic. It would be impossible to attempt any separation of these different registers: they belong together since the quest for oil, in Pasolini's Great Work, is as much poetic as geophysical or geopolitical.

In these Argonaut notes, short, fragmentary passages of Italian text are followed by parentheses that contain the words "Greek text" (*testo greco*). Now Pasolini could certainly have generated Greek text. In fact, elsewhere in the novel a similar construction is found that reads "Japanese text." And again, this reference to text to come acts as a pure presence for the reader. Pasolini writes at one point that what he is building is not a novel, but a "form" in the most ideal terms. After the Argonaut notes, he explains, he felt the need to explain the use of Greek (to come). Because the text is (not yet) legible, it is actually symbolic of this whole novel: "My decision is not to write a narrative, but to construct a form . . . a form that is simply consistent with 'something written.' I don't deny that certainly the best thing would have been to invent a whole alphabet, maybe with ideographic or hieroglyphic characters, for the entire work" (Pasolini, note 37, 155, translation mine). This language, unreadable to all but the author, would have most rigorously approximated a form without content, but as he goes on to explain, his character (shall we say his deeply entrenched humanism?) compelled him to avoid such extreme measures. So instead, he chose to use "apparently significant materials." However, *Petrolio* (with its notes in Greek as presence) must for now remain a form of discourse. As the author laments, it is not even a transitional object.

Petrolio remains unfinished. It is pure potentiality, or as Pasolini writes: "I am living the genesis of my book" (48). In this sense, Petrolio is intimately tied to fuels (the author claimed he conceived of the work after reading the word in a newspaper headline during the oil crisis) and also approaches—perhaps more than any work of fiction—what Lyotard also calls "a good book," that is, "a book where linguistic time (the time in which signification evolves, the time of reading) would itself be deconstructed—a book the reader could dip into anywhere, in any order: a book to be grazed" (Lyotard 2011, 13) In its unyielding writing-to-come, *Petrolio* is the greatest, most emblematic, modern work around fuel.

GEOTHERMAL (HEAT)

Ground waters received their heat early in the earth's formation when planets formed by gravitational accretion left behind molten material; or from radioactive decay of **thorium**, *potassium*, and **uranium** in the primordial interior of the earth. So inasmuch as "geothermal" can be considered a fuel, it is both ancient and mysterious. It can be used directly (district heating, for instance) or converted into electricity. In what may seem a perverse kind of anthropo-geological shift, Iceland (one of the youngest landmasses) is developing a program to export its geothermal heat beyond its borders.

Like many of his contemporaries, Cyrus Smith subscribes to a theory of entropic heat death—that is, one day the sun's energy will diminish to a point that life on earth will fade away. He puts forward another more idiosyncratic idea, though, about "the gradual extinction of our world's internal fires" (Verne 2001, 201). This is clearly what happened to the moon, he posits, and if God decides to enact it on earth the result will, at first, be massive migrations. "The Laplanders and Samoyeds will find the climatic conditions of the polar seas on the shores of the Mediterranean. And how can we assume that the equatorial regions will not prove too small to contain and feed all the people of the earth? Thus, why should nature, foreseeing this great migration, not even now be laying the foundation for a new continent in the area of the Equator, and why should she not have charged the [coral] infusoria with the task of constructing it?" (ibid.).

GOAT

Even on the subject of **gold**, Agricola's *De re metallica* is absolutely open, in stark contrast to the language of alchemists. Milling gold ores, for instance, can be achieved with **water**wheels, or with wheels tread by men or even goats. Their eyes should be covered by linen bands, he writes, although this detail is missing from the accompanying image. In the illustration for the process, the artist depicts a series of interconnected machines inside roofed structures. In the background is a landscape graced by a **wind**mill, which,

FIGURE 4. From Georg Agricola. Goats running together in an open and upright wheel, represent for our purposes something like an anti-alchemical form of discourse. *De re metallica*. Courtesy of the Columbia University Rare Book and Manuscript Library.

while not immediately part of the foreground machinery, is pleasing to the eye (see **wind** for more on this). What are we to make of the apparently "extraneous" decorations in some of the treatises' illustrations? Perhaps they are "tics" of the trade, disposable.[59] Certainly they lend charm. As Agricola explains, to counteract any ambiguity associated with the use of words (and particularly the veiled language of the alchemists), he hired illustrators to portray machines "lest descriptions which are conveyed by words should either not be understood by men of our times, or should cause difficulty to posterity in the same way as to us difficulty is often caused by many names which the Ancients have handed down to us without any explanation" (Agricola, xxx).

But inasmuch as the viewer lingers on them (rather than, say, making use of the book as a kind of reference work, and only turning to the particular section of interest in solving a particular metallurgic problem), they could also undermine the foreground, opening up new ways of thinking our relationships to the substances. This "subversive" mode of reading depends on recognition of the length and breadth or completeness of this project. At what point can we find cracks in this edifice of totality? Giving so much information about practices that are usually protected functions almost as an antidiscourse.

GOLD

Gold is not combustible. Still, it has the color and sheen of a flame, of heat. Gold may not serve literally as a fuel, but in writing there is no shortage of analogies in which gold warms the heart of a miser; or oil—liquid gold—is burned to warm the home of a millionaire.

George Eliot's Silas Marner weaves on a handloom, exchanging his caloric energy for gold—let us say a glowing perpetual cheery heap—which he hoards under the floorboards. He is warmed by the knowledge that it is there and he feels no compulsion to spend it.

Silas's cottage is centered around the hearth/focus, and he goes out only rarely. He has little need for human contact and, Eliot explains, his hoarding did not "beget any vice directly injurious to others" (64). Thus, Silas's hearth is a perfect figure of *oikos* in Lyotard's sense, as that which is secluded, comforting, free from public responsibilities, yet simultaneously tragic. It is

FIGURE 5. The men and beasts of Verne's Lincoln Island, untroubled by any exterior pressure to be part of a community, enjoy an intimate relationship with fuel. "Et le temps s'écoulait sans ennui" (The time passed without boredom), as the caption read in the context of *The Mysterious Island*. Courtesy of the Columbia University Rare Book and Manuscript Library.

only after the treasure is stolen that Silas is forced to look outside of himself: "[A]nd there was a slight stirring of expectation at the sign of his fellow-men, a faint consciousness of dependence on their good will" (125). Soon after the theft of his gold, a gift—the golden-haired infant he will name Eppie— arrives in his cottage, toddling "right up to the warm hearth, where there was a bright fire of logs and sticks" (170). On his hearth—the focus of his private space—he again finds gold (transmuted to the form of the golden-haired child).

While Silas insists on guarding the child as his, eventually, his loss and the subsequent gift will bring him into the community, and more specifically to attend church and to cease weaving on Sunday: "He had no distinct idea about the baptism and the churchgoing, except that Dolly [his only friend in the village] had said it was for the good of the child; and in this way, as the weeks grew to months, the child created fresh and fresh links between his life and the lives from which he had hitherto shrunk continually into narrower isolation" (197). Such community is a good thing for Silas, and inasmuch as this is a narrative of Christian redemption, we follow his movement with pleasure. Yet if we were to return to Silas Marner reading through the logic of fuel, we cannot forget that while his abandonment of the hearth (the place of fuel) in favor of the community represents social and historical progress, he has also moved to a place that is the place of the gift, expenditure, of *munus*, of *energeia*). He no longer has an intimate relation with fuel as matter. Is there some loss here worth considering?

GRAIL

Christ drank from the cup that becomes the grail.

In some traditions, it is (also? instead?) a glowing, energy-emitting stone (the philosopher's stone? Radium?) (Eco, 154).

In either form it fuels quests. Like the **fleece**.

GUNPOWDER

A mix of **charcoal**, sulfur, and potassium nitrate (saltpeter). Chinese alchemists may have discovered it in their search for an elixir to prolong life.

In 1680, Christian Huygens explodes gunpowder inside a cylinder to create hot gas that is released through valves. On cooling, the valves close down and a partial vacuum remains within the cylinder, creating steam that propels a piston up and down. Perhaps the first internal-combustion engine? For a brief moment, gunpowder acts as a fuel, before it is conscripted for other services.

HELIUM-3

A light, nonradioactive isotope of helium with two protons and one neutron. Although rare on earth, it is thought to be abundant on the moon.

In Duncan Jones's 2008 film *Moon*, helium-3, mined by a company called Lunar Industries, has solved the world's pollution problem (the term "climate change" is not used), along with poverty and hunger. We know this because the film opens with a commercial that informs viewers (of the film? Presumably earth-dwellers must be quite familiar with the marvelous effects of the salvation-fuel . . .) about life "now." We are shown older images of smokestacks and deserts, replaced "now" with smiling "ethnic" types holding apparently healthy babies in verdant fields. Our clean planet, brought to you by Lunar Industries, Inc. The infomercial raises many questions for the critical viewer: Does Lunar Industries have a monopoly? (We see only its apparatuses functioning on the moon). If so, why advertise? Is it a global corporation, or perhaps American, but one that has chosen to spread its technology to the entire globe out of charity, or because it is so easy to do so? And what about the energy infrastructures that take up the helium-3 and distribute it to the people? How do they function? Have they changed the modes of production of goods and services? How have they led to peace? Or is the fuel simply inserted into what was already there? But the viewer may not have time to formulate such questions adequately because the commercial ends abruptly and "the film" begins.

SPOILER ALERT: We will soon learn (long before *he* does) that the sole miner on the colony, Sam, is a clone. He receives occasional transmissions (tape delayed—supposedly because of a satellite malfunction) from a perky blond wife and a baby that he believes to be his. So we must ask: Is it possible that we are able to harvest helium-3 from the moon, to send it to earth in cryogenic pods to solve a wealth of problems, energetic and social? or to develop clones who do not know they are clones, a fully functional computer that controls the environment and the harvesters?—all of this, and yet we have not perfected something close to real-time communication from the moon to earth? So why would the clones—assuming they have been developed from a human prototype—accept the fact that they are cut off from their loved ones, laboring all alone for years with only reruns of *The Mary Tyler Moore Show* on a tiny screen to keep them occupied?

In his grim sets (reminiscent of the 1970s sci-fi he watched as a child), Jones has created a mining colony as dark and dismal as any coal mine of the past. Because lunar mining is automated, he can omit the large labor force and the sense of miserable collectivity that still conditions mining to this day. Yes, Sam is duped by the corporate entity that hires him. There is no security on the moon (although one would think that if the company can build an entire colony and transport fuel in three days, there would be a threat of breach by bad entities). The harvesters are named for the four Evangelists—no Islamic group in sight; no rogue states. The nuclear family is intact and technology on earth does not seem to have advanced in any significant way to disrupt social life. Indeed, as we hear very briefly in the radio transmission played over the final seconds of the film footage, the American government, appalled by the shady ethics of Lunar Industries (was there no regulation of this new fuel and its extraction?), has condemned the use of innocent clones. Sam is a hero for democracy.

Moon suggests a fantasy that if/when a fuel can come, so clean and powerful that it will profoundly change the earth, its extraction displaced in time and space, it will nevertheless not disturb us with profound questions about what it means to be human; what it means to live and labor.

HYDROGEN

Hydrogen—light, ubiquitous, carbon-free. "H_2 is the fuel [*sic*] of the future," we read in a brochure from Iceland, a lab of hydrogen. Hydrogen should not have an entry in this dictionary, except as a fantasy, since it is, more precisely, an energy carrier or storage medium. Moreover, as a gas it is too light to exist naturally. Almost all of world's supply is manufactured through processes like steam reforming of natural gas, gasifying coal, or partly oxidizing petroleum. In the reaction, which requires some other fuel, energy is released because negatively charged electrons in water are closer to positively charged nuclei toward which they are attracted than they were in the separate hydrogen and oxygen molecules. Once created, hydrogen can be stored for burning later. However, it is very volatile, so the process presents numerous logistical and chemical challenges.

In the 1920 and 1930s, hydrogen, released from zeppelins (where it helped to maintain buoyancy as in balloons), was recaptured to serve as a secondary or booster fuel. Engineers experimented with hydrogen in various forms of transportation in the decade that followed (Rifkin, 182).

Another fantasy: You are sitting in a 1959 De Soto Cella I, facing rearward and admiring the landscape through what looks like a Cinemascope screen— same aspect ratio, same sexy curves. You reach into the refrigeration compartment in the rear wall for a soda pop. You turn on a swing-into-position television receiver with "finger-tip controls." When you tire of watching television, you listen to taped music through stereophonic speakers in the roof. As you glide along the freeway, you are not thinking about fuel. Really, if pressed, you could not even identify the substance that is fueling your Cella I. Somewhere in the back of your mind, you know that the power source is an electrochemical cell, "based on new scientific developments, which would transform hydrogen and oxygen into silent electrical energy to drive four lightweight, high-speed motors located at the wheels." You also recall that "braking energy would be recaptured and stored in reserve batteries for future use, contributing fuel savings for tomorrow's motorist." You enjoy the "silent, silken operation . . . a flat interior deck, freed from transmission, differential and drive shaft humps."

In a fuel cell, such as one made in 1802 by Humphrey Davy, the fuel (which could be hydrogen, but also **methanol, ethanol, natural gas**, or **propane**) reacts with air rather than being burned in an internal-combustion engine.[60] Today, hydrogen, produced by the electrolysis of water (using renewable means), would be ideal for use in fuel cells, but with trace amounts of pollutants, and other energy issues related to transportation, hydrogen (or the "hydrogen economy") was probably never going to mitigate greenhouse gas emissions to a significant degree. And it is no longer the future that it was in the near past . . . Talk that is just a decade old, of distributed power, of the consumer of energy becoming her own producer, has faded, drowned out by the busy hum of the fossil-fuel industry.[61] And if it had been? The remnants of one dream: "When millions of small power plants are connected into vast energy webs, using the same architectural design principles and smart technologies that made possible the World Wide Web, people can share energy and sell it to one another—peer-to-peer energy sharing—and break the hold of giant energy power companies forever" (Rifkin, 9).

JATROPHA—SEE **BIOMASS**

JET—SEE **BITUMEN**

KEROSENE (OR PARAFFIN)

A clear liquid hydrocarbon obtained from the refining of petroleum. It was produced as early as the ninth century in a heated alembic, from shale oil or from **bitumen**. Kerosene was introduced as an alternative to **coal** oil (too smoky) or **whale** oil (increasingly too expensive) for indoor lighting. Abraham Gessner patented the name (kerosene = oil wax) and the method for distilling it in 1854. Kerosene served in heaters and stoves, and as a fuel for automobiles and even airplanes.

You are an American housewife, using a Florence-brand kerosene stove for cooking (endorsed by the Fuel Administration as a help in the conservation of coal for war purposes). It's a hot summer day, but you don't have to heat up your whole house just to make breakfast. Simply turn a lever to "light" and strike a match. Your burners (made from asbestos—ideal, clean, easy, a great conductor, an insulator, a kind of "antifuel" that does not burn!) help spread the flame around quickly. Then turn the lever to "burn." "Nor is there any smoke or odor, whether the flame is high or low, because there is no wick to get soggy, uneven and smelly."[62] So easy . . . so patriotic . . .

Because kerosene is a distillate of petroleum, it is not a satisfactory fuel if the problem we face is dwindling supplies or the perception thereof.

LEBENSKRAFT—SEE **VIS VIVA**

LIGHTNING

Many of Jules Verne's fantastical vehicles/narratives are powered by an unspecified *fée électricité*, futuristic bursts from electrical storms, strange charged flashes of St. Elmo's Fire, and so on. At times, "electricity itself becomes a principal protagonist in the drama" (Evans, 73).

And in *Robur the Conquerer*, the devious engineer, Robur, has built a flying machine that runs on "electricity" driven through rotating "screws" (propellers). His vessel, the Albatross, is threatened by a powerful electrical storm, but: "Robur, seizing the propitious moment, rushed to the central house and seized the levers. He turned on the currents from the piles no longer neutralized by the electric tension of the surrounding atmosphere. In a moment the screws had regained their normal speed and checked the descent; and the 'Albatross' remained at her slight elevation while her propellers drove her swiftly out of reach of the storm" (80). Eventually, the men on board are able to evade the "electric zone" and evaluate their situation: "The cause of this light must have been electricity; it could not be attributed to a bank of fish spawn, nor to a crowd of those animalculae that give phosphorescence to the sea, and this showed that the electrical tension of the atmosphere was considerable" (98).

At the 1939 World's Fair, General Electric set up a lightning demonstration meant to frighten and amaze, to remind families of the unthinkable forces that a company like GE could harness and drive into their homes.

LITHIUM—SEE DILITHIUM

LODESTONE—SEE MAGNET

MAGNET

The magnet possesses a **soul**, since it moves iron: Aristotle (citing Thales).

In his treatise on the subject, William Gilbert marvels:

> Wonderful is the loadstone shown in many experiments to be, and, as it were, animate. And this one eminent property is the same which the ancients held to be a soul in the heavens, in the globes, and in the stars, in sun and moon. For they deemed that not without a divine and animate nature could movements so diverse be produced, such vast bodies revolve in fixed times, or potencies so wonderful be infused into other bodies; whereby the whole world blooms with most beautiful diversity throughout this primary form of the globes themselves (308).

A young Descartes may have conceived of an **automaton**, powered not by hydraulics or air, but by a magnet. "Legend has it that he did build a beautiful blonde automaton named Francine, but she was discovered in her packing case onboard ship and dumped over the side by the captain in his horror of apparent witchcraft" (Price, 23). While this story is almost certainly false, it suggests that the force of a magnet might be greater than other mechanical means available, signaling a new paradigm to come in natural philosophy.

Christian Huygens wrote of a clock (invented by Hooke) that, moved by the power of a loadstone, and assuming it has no other motor, "would be a type of perpetual motion" and "an admirable invention" (cited in Coopersmith, 31).

In the early 1950s, the Banning Electrical Products Corp. showcased a concept car called Mota, powered by Polade—"the magnetic force of the universe." The brochure to accompany the exhibition explains that the Mota is "motivated by the magnetic forces which produce the speed of light." It "delivers flashing acceleration breath-taking speed, the pile-driving power of steam . . . all with the satin-smooth efficiency of constant horsepower electricity. With chain **lightning** harnessed beneath its bonnet, the MOTA's jet-flight ride hugs the highway with an exclusively new suspension system which automatically compensates for varying payloads while controlling vertical springing in both directions!" (Mota brochure). The brochure makes clear that magnetism (as in a Maglev, magnetic levitation train, for instance), by reducing friction, can make a vehicle highly efficient: "Reducing motive force to its simplest and most basic form, the MOTA engineering formula of Polade Power propulsion eliminates scores of working parts required by cars of standard design."

To recap: the car is related to or motivated by **magnet**ic forces, the power of steam, horsepower, electricity, **lightning**, and jets.

However, reading the fine print, we realize that the motor, mounted just above the rear wheels, is driven by an economical air-cooled gasoline engine-alternator combine that converts passive energy to positive energy at the wheels with the additional traction and acceleration of the increased torque. In other words, at a time of abundance, this car advertised space age, atomic, streamlined design (a drawing of an atom graced the cover of the brochure). **Oil**, not magnetism, fuels the vehicle . . .

MECHANIZATION

In the Highland Park Ford factory, production was fueled, precisely, by the desire to "save time, money, and manpower through further mechanization" (Douglas Brinkley, cited in Grandin, 35).

MEGODONT

Genetically engineered mammals chained to machines in Bacigalupi's *Windup Girl*. They have a trunk and four tusks (shaved off when they work in the factory). Ironically, while humans struggle for basic rights, they are represented by a union and a powerful one at that!

In addition to their living labor, deceased Megodonts can indirectly provide fuel. Offal that isn't eaten by pigs and Chinese ("yellow cards") is "dumped in the methane composters of the city along with the daily fruit rind and dung collections, to bake steadily into compost and gas and eventually light the city streets with the green-glow of approved-burn methane" (Bacigalupi, 22).

METHANE

Methane (**natural gas**) is a greenhouse gas, much more powerful than carbon dioxide in the short term, normally subjected to a process—the Brayton cycle—to convert it for use as energy. But before this, it lies in the ground, or under the permafrost; or it seeps insidiously into mines or disrupts the digging of major engineering projects; or it is produced by cattle or rice cultivation; through coal mining, and in the combustion processes that convert fossil fuels into thermal energy; by decomposing **waste** in landfills as in Bacigalupi's nightmarish Bangkok.[63]

The waste matter in landfills can be passed through phases of "putrefaction" and redemption, to invoke, again, the alchemical analogy, after which it is returned to the city as the "gold of energy." Outside São Paulo, in Badarantes, waste is compressed in many layers. Gas is extracted through tubes and carried to a power plant. "Scavengers have become alchemists who turn ordinary metals into a golden opportunity."[64] Former gleaners for scrap now transform matter.

"Purified" or transmuted landfill gas can be used to generate heat in a boiler, or inserted into an internal-combustion engine, gas turbine (electricity), or fuel cells (electricity), or converted into methyl alcohol and piped through existing natural gas lines.

Gas lit the streets of Victorian London. In Conan Doyle's "The Sign of the Four," Dr. Watson finds comfort in "the passing glimpse of a tranquil English home," whereas outside, on the "silent, gas-lit streets," the business of murder seems wild and dark. He is calmed, though, when he glimpses a **candle** in the window of number 3 Pinchin Lane, where Holmes has sent him to retrieve a dog with a refined nose (Conan Doyle, 116).

Venture capitalist Vinod Khosla has worked on a nearly closed-loop process (some corn is purchased from other farms) that uses methane from cow manure as a primary fuel. Now, for every BTU of energy used to run the plant, five are produced. This is much more efficient than a typical corn ethanol plant. An anaerobic thermal converter or Kergy machine produces cellulosic **ethanol** as a secondary fuel. There is no need for fermentation or acid hydrolysis, no need for organisms or enzymes to do work. Instead, organic matter—in theory any sort, including human waste—is heated in an oxygen-free environment to produce carbon monoxide and **hydrogen**.

A Mr. John D. Brown of Stueben County, New York, wrote to the head of Ford's Experimental Division, J. B. Dailey, on July 18th, 1916, on the possibility of substituting natural gas for gasoline in car engines. There would be various technical complications, Mr. Dailey replies, "but these would be trifling compared to the supply of natural gas . . . limited to about 15 or 20 years. At present we are devoting all our energies toward **alcohol** and its distillate from farm products, believing it to be the proper substitute for gasoline"[!]. At the present moment, especially given new extraction techniques such as hydrofracturing (fracking), the earth's supply of natural gas seems almost incalculable.[65]

Despite numerous advertisements of the contemporary infosphere that label gas as "clean," "green," or "patriotic," we should recall that it is, first and foremost, a fossil fuel. Natural gas, a mix of hydrocarbons (mostly methane, but also ethane, propane, and other gases) can be drilled from the ground or produced as a by-product of crude-oil refining. To be sure, compared with coal, natural gas produces much less sulphur and particulate in burning. Overall, the carbon emissions are also lower.[66] It can be stored in a compressed gas state or liquefied (lng). It is primarily delivered through pipelines and more evenly distributed geographically than oil. Whereas oil requires rather precise and unusual conditions to develop, natural gas has fewer geological requirements. John McPhee makes use of an extended *analogy* to describe it: "Natural gas is to oil as politicians are to statesmen" (179). Poor methane, always an ambassador to some godforsaken country with a pipeline to somewhere strategically significant, never a cabinet member. It lives for a relatively short time (twelve years) in the atmosphere compared with vastly longer horizons for CO_2. But the former could warm the planet a hundred times more than the later during that period. Perhaps this dictionary should not trouble itself with the effects of leakages during the process of extraction. We just need careful regulation. We need the jobs. It's a bridge fuel. Just a little longer. A few more years. Another election cycle and then . . .

METHANE HYDRATE

A mixture of methane and ice, it is found in the (melting?) permafrost of the Far North and on the ocean floor, along seismic fault lines. There are probably vast wells of methane hydrate available as the permafrost melts, and while at the moment it will probably require more energy to extract it than it would provide, more may become easily accessible.[67]

Yet, let us repeat: methane is a very powerful greenhouse gas.

METHANOL

An alcohol in flex-fuel vehicles (M85- = 15 percent gasoline), methanol enhances octane. Because it contaminates groundwater, it is not used much. Moreover, it is not, for our purposes, a "future fuel" but one that may be used to increase the efficiency of and decrease the greenhouse gas emissions of more carbon-intensive fuels, soberly, in the present.

NATURAL GAS—SEE METHANE

NITROGLYCERINE

Explosive liquid with medicinal applications, commonly made by treating glycerol (a colorless liquid) with nitric acid (a by-product of combustion; **lightning**).

A man writes to Ford in 1916 suggesting the use of nitroglycerine for use as an automobile fuel. Yes, there is a risk of explosion, he admits, but not if the acids and glycerine are kept separate. "My idea is to have separate tanks, and have a small drop mix only when it reaches the cylinder."[68]

The response, by J. B. Dailey, Ford's chief engineer: "I wish to advise you that we are afraid you are working along the wrong lines for a substitute for gasoline inasmuch as the ingredients such as nitrate and sulfuric acid with gasoline as a binder would cost too much and do not think you could compete in price with gasoline."

OCEAN

Oscillating water columns filled with air could be placed in the sea. When hit with waves, the air might be compressed and then forced into turbines to generate electricity.[69]

In the 1840s, James Thomson (brother of William) developed the idea of a tidal mill, a kind of **perpetual motion** machine to grind corn, using power stored "in the motions of the earth or moon or one of them," but a fellow scientist denied the possibility that water could "deduct" from tidal motions (Smith, 34).

The *Nautilus* of *Twenty Thousand Leagues under the Sea* becomes stuck in a reef. Captain Nemo, though, is unfazed. Why worry when the tides will eventually do the hard work of dislodging the submarine from its present position? Of course he is correct, much to the amazement of his captives.

Ocean ways, while unpredictable, could exercise considerable force as a fuel. They might be increased in motive power by superstorms and other effects of climate change. Superstorm superpower surges?

OIL . . . (OR AS UPTON SINCLAIR WRITES: OIL!)

At the time of the first oil well, drilled in 1859 in Titusville, Pennsylvania, oil was not used as a fuel for transportation and industry, but rather, distilled into paraffin (**kerosene**—for illumination) and other compounds. It was also marketed as a healthy elixir to be drunk.[70]

In order to really think oil as first fuel, it seems important to peer down into the wells, before the material is sent to a refinery and before it is inserted into a text. To gaze down through many layers of compressed matter into deep time. Of course, we can't really do this, except through a cutaway view, something very artificial. Perhaps oil is less narratizable than coal, as Michael Ziser has suggested, precisely because it spreads and gushes; it is less localizable.[71] In an influential essay first published in *The New Republic*, Amitav Ghosh wondered why there was so little fiction about oil. In the American context, he suggests, "oil smells bad. It reeks of unavoidable overseas entanglements, a worrisome foreign dependency, economic uncertainty, risky and expensive military enterprises. . . . It reeks, it stinks, it becomes a Problem that can be written about only in the language of Solutions" (Ghosh, 139). Indeed, he posits, oil and the novel are at odds because the places of oil are "bafflingly multilingual," while the novel is more suited to "monolingual speech communities" (142). In contrast, Reza Negarastani's *Cyclonopedia* would lead us to believe that oil—today—is not only the lubricant of narrative, it is narrative, the only narrative, all narratives. "Events are configured by the superconductivity of oil and global petro-dynamic currents to such an extent that the progression and emergence of events may be influenced more by petroleum than by time. If narrative development, the unfolding of events in a narration, implies the progression of chronological time, for contemporary planetary formations, history and its progression is determined by the influx and outflow of petroleum" (Negarestani, 22).

Go back a few decades, to the period of the so-called Middle Eastern oil crises. Although Pasolini's *Petrolio* is more than five hundred pages long, it does not engage with oil as matter; no oil is spilled along with the ink to make up this epic.

Before him, here is Upton Sinclair:

> The greatest oil strike in the history of Southern California, the Prospect Hill field! The inside of the earth seemed to burst out through that hole; a roaring and rushing, as Niagara, and a black column shot up into the air, two hundred feet, two hundred and fifty—no one could say for sure—and came thundering down to earth as a mass of thick, black, slimy, slippery fluid. . . . It filled the sump-hole, and poured over, like a sauce-pan boiling too fast, and went streaming down the hillside. Carried by the wind, a curtain of black mist, it sprayed the Culver homestead, turning it black, and sending the women of the household flying across the cabbage-fields. Afterwards it was told with Homeric laughter how these women had been heard to lament the destruction of their clothing and their window-curtains by this million-dollar flood of "black **gold**"!
>
> Soon the streets were black with a solid line of motor-cars. (Sinclair 2007, 25)

Nevertheless, aside from a few similar passages in which "Dad" and "Bunny" see, feel, smell, and touch the oil that spurts up in Southern California, the bulk of the prose of Sinclair's novel is dedicated to the social life of Bunny, his education—political, sexual—during Prohibition (no booze in the gas tank in Southern California), Hollywood, the parallels lives of the financial and petromarkets, and so on.

The whole culture of oil is so new in Sinclair that terms like "filling station" and "gas" are placed in quotation marks along with modern design principles ("built-in" features of a home) and, of course, fuel figures ("black gold").[72] At the end of the novel, Bunny's friend Ruth falls into an oil well, finding her death underground like a series of literary characters from epic and narrative, and more locally, like one of the first workers to operate on Dad and Bunny's land. Her corpse does not emerge preserved, like a **peat** bog body or the miner of Falun.[73] Nor is she saturated with fuel like one of Zola's colliers. Rather, the hole is empty because the Southern California oil is rather shallow and life will now begin to flourish on the surface as Bunny and his bride, Rachel, found an academy for socialist thought. From now on, we might say, aside from a few tired pumps that continue to operate to this day, the Southern California landscape will grow verdant and Technicolor; and gas will be held, waiting, in tanks, for motorists who will have minimal contact with its smell, its viscosity. Oil will be "other" and displaced for the greatest car culture in the history of the world.

You are driving along the highway in postwar Italy. You have forgotten to fill up the tank (of your Fiat? 500? 600? Or if you have really moved up the consumer chain a 1100?). It's the lunch break and the filling stations are closed. You remark on this fact and then you note: "Every year two million tons of crude are brought up from the earth's crust where they have been stored for millions of centuries in the folds of rocks buried beneath layers of sand and clay" (Calvino, 170). You slip into this prose with the ease of oil flowing from the pump itself. There is no break in the flow of language: your situatedness in a particular place and time and your description of the subterranean and its time are perfectly fused. As you drive, you think about the fuel gauge in your car and simultaneously about the global geological fuel gauge. You say to yourself: "My foot on the accelerator grows conscious of the fact that its slightest pressure can burn up the last squirts of energy our planet has stored" (171). There is no reason you should distinguish here—in this context—between oil as fuel and oil as (stored) energy. This distinction is not important for your motion forward. On the contrary, thinking critically while driving might create hazardous conditions, especially on these roads filled with the inexperienced drivers of a boom culture.

Bernardo Bertolucci makes a television documentary (sponsored by ENI, the Italian state hydrocarbon company) called *The Petroleum Route* (*La via di petrolio*). His camera lingers on laughing Persian children (the film is dedicated to them), the blue stones of Isfahan mosques, the tankers

that chug slowly across the sea, and the immense pipeline that carries refined gas from Genoa up to Inglestad. There are no close-ups of oil. Oil does
not enter into our senses, and the narrator even reminds us that when we
fill up at the pump, we do not think about the sacrifices and dangers of
extraction.[74]

There is one crucial moment in the documentary, though, in which Bertolucci creates, through editing, an indelible link between the human and
oil: it is when he intercuts footage of Iranian men bobbing up and down at
prayer with footage of the desert pumps making the same movements. The
effect is absolutely uncanny.

PATRIOTISM

The men of Verne's "Lincoln Island" are so accustomed to their new life,
so perfectly at home, that they almost secretly fear it will come to an end.
Still, Verne admits, one fuel might have propelled them from their colony:
"So strong is the love of country in the hearts of men that if some ship had
unexpectedly appeared on the horizon, the colonists would not have hesi-
tated to signal it, attract it to them, and sail away!" (Verne 2001, 306).

PEAT (SEE ALSO BIOMASS AND COAL)

Peat is **coal**, but coal that is much younger than what we burn, allied more
readily with the human capacity to imagine time. "Peat bears much the same
relation to coal that snow does to glacier ice. As snow is ever more buried
and compacted, subjected to geothermal heat, it gradually gives up much of
its oxygen, hydrogen, and nitrogen, and concentrates its content of carbon"
(McPhee, 247).

The primary source of fuel during the Dutch Golden Age, peat is slightly
buried in boggy wetlands and has thus not been oxidized. Although it must
be dug out—no easy task—because peat lies close to the surface, the cost
of transport to consumers via water was quite inexpensive. It required little or no capital investment or organization, so local farmers were able to
produce peat for heating to meet the rising demand in towns and cities on
a rather informal basis. A new tool, the *baggerbeugel*, facilitated cutting
below the surface. Bogs quickly began to develop into lakes, and the government stepped in to regulate and reclaim some lands. Eventually, other forms

of peat were discovered in alluvial zones, requiring a greater capital input, as demand for export rose. Canals were dug to convey the peat, changing the whole geography of the Netherlands with tremendous implications for the present of rapidly rising sea levels. In any case, as a fuel, peat is short-lived, nonsustainable. And that is considered a key reason why the early modern Netherlands did not progress into advanced industry but remained for a long period a state of mercantile wealth.[75]

Evelyn praises the Dutch, who lack fuel (that is, **wood**) but who make do with peat "eggs" that burn efficiently in a particular type of oven (that is, they function within a larger energy system). "The right mingling, and making up of charcoal-dust . . . there is not a more sweet, lasting and beautiful fuel."
He goes on:

> The manner of it is thus: Take about one third part of the smallest of any coal, pit sea, or char-coal, and commix them very well with loam (whereof there is in some places to be found a sort somewhat more combustible) make these up into balls (moistned with a little urine of man or beast) as big as an ordinary goose egg, or somewhat bigger; or if you will in any other form, like brick-bats, &c. expose these in the air till they are thoroughly dry; they will be built into the most orderly fires you can imagine, burn very clear, give a wonderful heat, and continue a very long time. (2:107)

Peat is, today, an extremely valuable store of carbon, like rain forests. In England, though, peat is also an essential "fuel" for successful home gardens. Some environmentalists refer to peat harvesting as "mining" and oppose its use both in terms of the disturbance of local ecosystems (bird habitats are located in the bogs, for instance) and in the context of climate change (carbon emissions). Because bogs grow very slowly (less than half an inch in a century), they are not considered "renewable." In England, the gardener demanding peat at the local nursery finds herself on the wrong side of the carbon debate.

PERPETUAL MOTION

Not one fuel, but a dream of a fuel, not one but many, or rather, one that is multiple. Like **air**, perpetual motion represents pure hope: free, unlimited, provided by Nature with no apparent negative collateral effects, a deferral of fuel to somewhere else in space and time.

A Sanskrit text instructs how to create an early perpetual motion machine, an overbalancing device (a rotating wheel also symbolized eternal cycles of life):

> Make a wheel of light timber, with uniformly hollow spokes at equal intervals. Fill each spoke up to half with mercury and seal its opening situated in the rim. Set up the wheel so that its axle rests horizontally on two supports. Then the mercury runs upwards [in some] hollow spaces and downwards [in some others, as a result of which] the wheel rotates automatically forever. (cited in Coopersmith, 7)

Foucault's pendulum (also the place of the dramatic and alchemical conclusion, a "hermetico-energetic" ritual of Umberto Eco's novel of the same title) could, without obstacles, oscillate from pole to pole for eternity.[76] Fueled by the forces of the earth's rotation.

In automotive technology, a flywheel needs an initial push but then rotates for a relatively long time. As with the **automaton**, a viewer who encounters it at any given point in its cycle might mistake it for a perpetual motion machine. Only one who was present at the start-up would see through its mask.

A mechanic writes to Ford to announce that he has invented a perpetual motor powered by "dead weight" that could move a machine twenty-four feet in diameter with unlimited horsepower. As is typical of many men in the first half of the twentieth century, he feels a direct connection with the industrialist. He begs: "If you will help me complete it we will share the glory of this invention" (cited in Wik, 75).

Indeed, when it first came on the scene, **radium** was seen as a form of perpetual motion. Here is Frederick Soddy: "The driving power of the machinery of the modern world is often mysterious, but the laws of energy state that nothing goes by itself, and our experience, in spite of all the perpetual motion machines which inventors have claimed to have constructed, bore this doctrine out, until we came face to face with radium. Nothing goes by itself in Nature, except apparently radium and the radioactive substances" (Soddy, 21).

His colleague, Ernest Rutherford, recounts an experiment that is as close to a perpetual motion machine as he has ever witnessed: A scientist has placed radium in a glass container along with gold leaf. B-rays emitted from the sample carry away negative electricity. This causes the gold leaf to peel off a little at a time until it touches the side of the vessel. There is then a discharge that Rutherford believes might last for a thousand years, although with ever diminishing power. Recall that for Rutherford, this "machine" comes prior to "transmutation" (a force imagined to produce enormous potential energy) that he predicts (correctly) will be achieved in the future.

A MINI-DRAMA, c. 1816:

Robert Stirling: "There are plenty of theoretical perpetual motions if we have friction, resistances &c out of consideration."
James Stirling: "Yes, but not perpetual sources of power."
Pause. Silence. (Smith, 50)

Much earlier, sober minds realize that the dream of perpetual motion is impossible. William Gilbert, for instance, dismisses the physics of perpetual motion machines:

Cardan writes that out of iron and loadstone may be constructed a perpetual-motion engine—not that he saw such a machine ever; he merely offers the idea as an opinion. . . . But the contrivers of such machines have but little practice in magnetic experiments. For no magnetic attraction can be greater (whatever art, whatever form of instrument you employ) than the force of retention; and objects that are conjoined, and that are near, are held with greater force than objects solicited and set in motion are made to move . . .

And without mincing words he goes on to deride those who, in the distant past, repeated the false dream:

Such an engine Petrus Peregrinus, centuries ago, either devised or delineated after he had got the idea from others; and Joannes Taysner published this, illustrating it with wretched figures, and copying word for word the theory of it. May the gods damn all such sham, pilfered, distorted works. (166)

Petroleum—See Oil

Philosopher's Stone

Some alchemists believed that the stone—whether procured or produced—could act as fuel to transform lead into **gold**. But for others, the stone was itself the end product of this initial transmutation. The history of alchemical writing is heterogeneous, and many different traditions vie for authority on questions such as this one. Let us say, then, that the stone is an unstable signifier of successful transmutation. We have seen that in general alchemy raises for us the problem of "prime matters." Is the prime matter of alchemy the philosopher's stone—a material revealed to only a select few? Either the stone is a *prima materia*, a *massa informis* that underlies a process of perpetual change, or something closer to a raw material, provided by nature but worked by an artisan or chemist and thus available—as potentiality—for a secondary stage of transmutation.

Flamel might have been able to achieve the Great Work, but he lacks what he calls "the stone," implying that it is not the final product, or even the sole origin, but an element of the process. Apparently the book by Abraham the Jew (not extant, perhaps wholly invented) suggests that the stone requires cabalistic knowledge, conventionally passed down from father to son, or perhaps learned in a particular context. So Flamel travels to Spain to find a Jew. And although the (converted!) Jew dies on the way back to Paris (he is duly buried in a churchyard), Flamel has learned how to prepare the stone, which is not apparently some matter already existing as such. He writes: "I found that which I desired, which I also soone knew by the strong sent and odour thereof. Having this, I easily accomplished the Mastery, for knowing the preparation of the first Agents, and after following my Booke according to the letter, I could not have missed it, though I would" (Flamel, 28). His first projection, on mercury turned into silver "better than that of any Mine." Then he is able to produce gold. He achieves transmutation three times—precise dates and times are given, as if to add verisimilitude—and he and his wife, Perenelle, donate their riches to charitable organizations.

In his *Sylva*, Evelyn makes an analogy between gold and wood, or better, a comparison: both are precious, but "we had better be without gold, than without timber." And yet it is not of "universal use 'till it be duly prepar'd"

(2:80). Like gold, wood must undergo a process (seasoning) of transformation. Its value beforehand is in question. Does this extend to the stone itself? It is easy to imagine how modern scientists working with radioactivity found an analogy with alchemy in general (and with the stone in particular). For the stone emerges as a kind of fuel (secret or not) for a secondary set of processes. There exist other alchemical narratives, however, which end—abruptly—when the stone has been made. Such we might say are the equivalent of tales of drilling down an exploratory well that ends in a spurt of oil and a cry of "Eureka!"

Ideally, alchemy would produce gold from something of little or no value, yet some early modern alchemists advocated the use of precious metals (albeit in small quantities) as the elements that would be projected, purified, and expanded to produce the philosopher's stone and eventually high-quality pure gold itself. Again, radioactivity provides an analogy for this potential paradox. Soddy writes that to build up a heavy element like gold from a lighter element like silver would require, in theory, a huge expenditure of energy. In such a case, then, "that ounce of gold would be dearly bought." However, he goes on: "On the other hand, if it were possible to artificially disintegrate an element with a heavier atom than gold and produce gold from it, so great an amount of energy would probably be evolved that the gold in comparison would be of little account. The energy would be far more valuable than gold" (Soddy, 173). If we extend alchemy as a model from nuclear physics to various green "future" fuels, the question of prime matter again becomes significant. Biofuels, for instance, are often criticized because the production of the "prime matter" may take away from foodstuffs, and its transformation into energy requires energy. One need only think of the debates—raging in the very recent past, toned down a bit since the 2012 Midwest drought and the booming of shale gas—about corn and **ethanol**, for instance.

In his early article for the *Reinische Zeitung* on fuel (the customary right to wood), Marx addresses "what the learned and docile servants of the so-called historians found to be the true philosopher's stone which could form any impure pretension into the pure gold of right" (Marx 1977, 26). Let us clarify some of the terms in this analogy: the base matter here is the customary law by which the poor were able to gather dead wood from the forest floor. The gold is modern, bourgeois Law, and the stone is a mode of writing of history that serves to naturalize Law. It is a kind of illusion, but a real illusion. Marx calls for something like a reverse transmutation, back to the base; that is, a movement back to "the right of custom" that belongs to the property-less poor.

As in early modern alchemical writings, so in thought of energy in the eighteenth century (but for entirely divergent reasons), "reverse transmutation" is unnatural and abhorrent. "Everything in the material world is progressive. The material world could not come back to any previous state without a violation of the laws which have been manifested to man, that is, without a creative act or an act possessing similar power" (Smith, 110). Like many of his colleagues, the Glasgow scientist William Thomson espoused a theory of entropy or solar or heat death (the physical world has a limited time span of survival). He further believed that man himself had no power to reverse the course of energies. Still, one could certainly "direct [those] energies to beneficial and productive ends such as the driving of a vortex turbine from a fall of water which, though in itself aesthetically pleasing, would otherwise be wasted to mankind" (cited in Smith, 111).

PLUTONIUM

A heavy and rare radioactive element, plutonium comes from the ground, hence its name is linked to the god of the underworld. It can be produced, like the philosopher's stone, as a second matter, during nuclear fission (from uranium-238 nuclei). Plutonium fueled the bombs used at the Trinity test site in New Mexico and in Nagasaki. It was also tested on human subjects.

Since the early 1960s, plutonium has fueled manned but, more commonly, unmanned space missions. The Mars Curiosity rover that landed in 2012 after a complex set of maneuvers has a rod of plutonium 238 that can be converted into heat (to keep the terrestrial mechanisms performing in the cold environment) and electricity to perform its functions. NASA officials routinely lament that so little of this precious fuel is available. Unless a suitable replacement can be found, future exploration of space may be curtailed, they say.

In the 2013 film *Oblivion*, plutonium powers a ship that crashes on a deserted earth. Conveniently, it's weapons grade: Jack Harper (played by Tom Cruise) and the leader of the human band of survivors (played by Morgan

Freeman) will repurpose the fuel to explode inside an enormous vessel called the Tet that has been sucking up earth's seawater as fuel. For more on this, see **water**.

POTATOES (SEE ALSO **ALCOHOL**)

Letter from William Carter, Toronto, November 24, 1916, to J. B. Dailey of the Ford Motor Company offering potatoes grown on his land in Manitoba for fuel. He has been in contact with "the highest Chemical and Agricultural authorities" and with "the Government," but they have not responded. He pleads his case directly to Henry Ford: "I am willing to make any reasonable arrangement to get the enterprise started, but although I own the land and also property in this city and also in your country I am not capitalist. I would rent my land at a nominal figure with the option of purchase."[77]

PROPANE

In the United States, propane is used in rural areas and to fuel backyard barbecues. It's booming along with natural gas. Lpg is mainly propane mixed with other hydrocarbons. It is a by-product of natural gas processing and crude-oil refining, with a global market.

In a recent brochure promoting the gas, a human hand holds a plant seedling. In the text propane (a fuel) is proposed as an alternative to electricity (an energy system or force). The brochure goes on to explain the image: because power plants that produce electricity pollute the air, propane is clean(er). Than . . . ?

Why should a local rural energy provider (or Jules Verne, to recall his muddled language around electricity) be concerned to locate and separate fuel as matter distinct from or prior to a system into which it is inserted? They would have no motivation to do so either because we are more compliant consumers of energy if we do not think about fuel; or because we are more engaged readers of scientific novels if the author preserves some mysteries.[78]

RADIUM

"Discovered" in 1898 by Marie Curie, radium is highly radioactive (over a million times more so than **uranium** of the same mass). The naming of radium dates to soon after its "invention," from Latin *radius* (*ray*), so called for its power of emitting energy in the form of rays. Because of its instability, radium glows with a faint blue light; hence it captured the imagination of the public more than other radioactive elements. Radium decays in stages (analogous to the alchemical process), yielding radon (a gas) and then a series of solids: radium A (polonium, named for Curie's country of origin, Poland), radium B (lead), radium C (**bismuth**), and so on. Each of these products is radioactive. Indeed, radium itself is a product of decay of heavier elements (mostly uranium ores). It is very rare, and found in only minuscule quantities; hence it is not used as a nuclear fuel.

Radium was considered to have life-giving properties and was associated with heat, **perpetual motion**. Soddy notes that the radioactivity produces four basic effects, a number of which only became visible with radium. First, radioactive substances can affect a photographic plate; they can excite phosphorescence or fluorescence in other substances; they cause air and gases to lose insulating power they normally possess and become conductors of electricity. So any electrified object has its electricity rapidly discharged when placed near a radioactive substance, and the same could be said of x-rays, incandescent bodies, and lighted matches. Radioactive bodies generate heat like coal or any other combustible material. But Soddy notes that "in the naturally occurring radioactive substances these effects are far too small to be readily detectable" (9). With **uranium**, for instance, it was not clear that radioactive elements actually evolve a store of energy "presumably out of nothing" (10). Soddy explains: "It does not require much effort of the imagination to see in energy the life of the physical universe, and the key to the primary fountains of the physical life of the universe to-day is known to be transmutation." If radium can be harnessed (for transmutation), he predicts, we might fuel travel to outer space and make the climate temperate: "We could transfer a desert continent, thaw the frozen poles [!], and make the whole world one smiling garden of eden" (182).

When he realized the potential for radium to act as a fuel, Rutherford is said to have warned his partner: "For Mike's sake, Soddy, don't call it transmutation. They'll have our heads off as alchemists" (cited in Campos, 3). But as Soddy notes, the analogy is not (merely) "fanciful." That is to say, figuration can and must be tolerated in this context because the connections between alchemical transformation and radioactivity are so empirically intimate, proximate.

How far does the analogy extend? If alchemy was, in the early modern period, both practicable (production of noble metals in a laboratory) and spiritual (production of a form of redemption), or better, practical (allied with various early forms of chemistry, apothecary mixing, metallurgy) and theoretical (never actually achieved, for this would signify a form of base greed), can we also expect to locate ambivalence in relation to fuel? Or does the alchemical analogy go only as far as the beginning of the process and then fall apart? If so, then alchemy is like a fuel in that it stands as legitimate first matter *in potentia*, but as soon as it is "processed" it is stripped of its power. It is a figure that could (and most often does) become meta-phorized, but as soon as it is transported, it loses its power to signify.

And then, "when was" alchemy? From the point of view of fuels, perhaps it was prior to modern chemistry, for instance, prior to the periodic table. So is the status of all "alchemy" that comes afterward merely analogical with a taint of the historical? In his lectures on radium titled "The Newer Alchemy," Rutherford attributes the origins of the term to the Early Christian era, and its flourishing in the Middle Ages, thus ignoring a whole series of ancient texts and traditions (not to mention the seventeenth century, alchemy's "golden age"). In the end, for Rutherford, the alchemists of the Middle (or any) Ages never did change lead into gold except in their writings. Rutherford understands that alchemists searched for (literally? experimentally?) a **philosopher's stone** that could transmute one element into another. Aristotle helped the alchemists believe that "all bodies were supposed to be formed of the same primordial substance, and the four elements, earth, air, fire and water, differed from one another only in possessing to different degrees the qualities of cold, wet, warm and dry" (Rutherford, 1). He goes on to explain that transmutation was never actually achieved until his present. So in his unambiguous introduction to his rather technical lectures (in contrast with Soddy, or with H. G. Wells's scientist, who perform radioactivity for the public), there is a precise historical and rhetorical status to the alchemical: it was a theory of matter based on Authority, tested without success in Early Christianity and proven only in the modern period. The link between the stone and radium could not be less figurative.

Indeed, after the introduction, Rutherford's little book never mentions alchemy (or history, or figures) again, until the final paragraph. So far, concludes the physicist, we have been able to study transformation by bombarding elements with particles or with slow neutrons. The reader expects a prediction of transmutation (as we find in Soddy, to be sure). Instead, at the end of the book, alchemy ceases to serve as an analogy for a process-to-come, a transformation leading to a "future fuel." Instead, it is, quite simply, a rather impoverished, exhausted metaphor for knowledge of the elements themselves, qualified by an adjective ("modern," to distinguish it from "ancient") and attached to a notion of progress (although it is unclear from the prose if alchemy is an [allegorical?] progress, or if it has progressed since its origins). And Rutherford concludes, "Thus [we] will see how the progress of modern alchemy not only adds greatly to our knowledge of the elements, but also of their relative abundance in our universe" (67).

In *The World Set Free*, H. G. Wells images a professor (a figure obviously based on Rutherford) who gives public lectures on radium. The professor explains the uniqueness of his mineral sample, not in quality but in terms of scale. He tells a rapt audience: "It does noticeably and forcibly what probably all the other elements are doing with an imperceptible slowness. . . . Radium is an element that is breaking up and flying to pieces. But perhaps all elements are that at less perceptible rates." And he goes on: "We know now that the atom, that once we thought hard and impenetrable, and indivisible and final and—life-less—lifeless, is really a reservoir of immense energy" (21). Following the most contemporary science of his day, Wells's professor surmises that **uranium** gives off radium, then radium changes into a gas called the radium emanation, and then into radium-A, giving out energy at every stage until it reaches its final stage, presumed in the narrative time of the lecture to be lead. In other words, in the early days of radium experimentation, suffused with alchemical rhetoric, the entire process mimics reverse transmutation, a return of the noble to the base that threatens to undo the Great Work, indeed to undo narrative itself.

Although it is difficult to generalize about a very broad corpus of texts, most alchemical writing assumes, following Aristotelian thought (as per Rutherford), that a material substratum persists beneath all accidental and substantial changes in nature. Human craftsmen do not create anything from scratch. Rather, they take some preexisting bit of matter at hand and imprint a form on it. In this sense, all matter is pure potentiality or *dynamis*. It cannot be without importance that the term for matter is the same for wood (fuel)—*hyle*.

Yet the prime matter itself does not possess any particular ability. It is not potential or potential energy, which also means that it can never be corrupted. For instance, the Renaissance philosopher Marsilio Ficino understood matter as kindling (fuel!) with the form of all things, and in formless prime matter "certain seeds of forms lie hidden and ferment." "When he uses expressions like these Ficino is generally trying to capture the cracking and latent power that he thinks belongs to both prime and corporal matter: the imperceptible sparks of all things that kindle in matter await the fanning of soul in order to issue forth from potency into act" (Snyder, 217).

In the "present" of Wells's novel, when radium has been harnessed as fuel for transportation, science has yielded another sort of anxiety. As radium disintegrates into a heavy gas and is consumed, it turns out that it yields not lead, but **gold** as a waste product, eventually leading to the downfall of the world's economy.[79]

In various early modern alchemical texts, we find a number of strategies to combat or disavow this kind of threat. Either the alchemist does achieve the Great Work, but the product is **gold** in such minute proportions that it cannot devalue markets, or the product is a spiritual one for which gold stands as a mere figure; or, in the case of Flamel and Perenelle, for instance, they donate their wealth (achieved thanks to the wisdom of a Jew, it should be recalled) rather than using it to purchase goods or raise their social status. At least this is the narrative constructed around them.

In *The World Set Free*, then, the excess of **gold** as a waste by-product of radium cannot be taken up or recycled in any useful way since gold is already so deeply charged with meaning in the culture.

SALT

Salt keeps the sacrificial victim from putrefying. It slows the process of decay, functioning perhaps as an antifuel.

Osmotic power, created by the separating of salt and fresh water, has been discussed in the portfolio of alternative forms of energy. In this case, salt would not be a first fuel, only a catalyst in a process.

A W. E. Mortrude Jr. from Seattle writes to Henry Ford in 1916, proposing a special engine called the Oxydrogen Internal Combustion Engine that runs on a cheap and widely available fuel: salt.[80] A by-product of Mortrude's process, caustic soda, could be made into soap. Mr. Mortrude has already proposed his engine to Navy Secretary Daniels, but without satisfactory response. So from Ford, he is willing to accept "most any reasonable proposition that you may present. To place me on a salery [sic] to perfect it, and give me a small interest, or most any way that you think advisable." In a follow-up, Mortrude clarifies that he has also developed a special process for the production of salt. And there is another by-product of the process worth mentioning: oxygen. This could grow to be "an immense seller." Finally, he adds that he has a plan for transporting the salt from the factory site: A "high class **wind**mill." He notes, "The fuel must be transported anyway, no matter in what form, whether it is in the form of sodium chloride or in the finished product as metallic sodium and sodium dioxide in their respective cells. For convenience, if the Ford Motor Co., had a large Aero-Electric plant in Kansas, or several of them the salt could be transported to the plants in said state and the byproducts could be disposed of as nearly as possible within the state boundaries, the fuel as well, of course."

A response indicates that Mortrude's letters were placed on Mr. Ford's desk for his consideration. Did Ford read them? And if so, did he think about a total system (a purpose-built engine + a patented process of fuel cells + a renewable power source for transportation of fuels + disposal of by-products)?

In *Twenty Thousand Leagues under the Sea*, Nemo has found a way to harness osmotic power:

One way would be to establish a circuit between wires set at different depths [of the ocean] and get electricity by means of the reaction to different temperatures "sensed" by those wires. But I prefer a more practical method.... Seawater is water and sodium chloride. I extract sodium from seawater.... Mixed with mercury, sodium forms an amalgam that takes the place of zinc in Bunsen batteries. The mercury is never consumed, only the sodium is used up, and the sea resupplies me with that. (Verne 1993, 78)

The professor on board is duly impressed: "Marvelous! I agree, captain, you're right to exploit this source of energy. It's sure to take the place of wind, water and steam" (81). Yet—perhaps because he has already explored it and must then move on to other forms—by the time we get to *The Mysterious Island*, this form of power has been abandoned in favor of others . . . including . . .

SEAL—SEE **TALLOW**

SEAWEED—SEE **ALGAE**

SOUL

Some early thinkers classed the soul as a source of motion (that is, possibly, a fuel). Diogenes, for instance, believed the soul was comprised of **air**, a first principle capable of originating movement. In Allan Stoekl's Battaillian vision of a post-sustainable, post-fossil-fuel future, the soul has ceased to be a possible fuel. There is no soul. Only an economy of excessive bodily expenditure: gleaning, recycling, walking and cycling in the city. Man's freedom comes not from the soul but from the body as pure excess. In the most practical terms, this future could be one of extreme hardship, sacrifice, austerity. Bataille opens a way to think . . . otherwise.

STAKHANOVISM

A superhuman desire to succeed for the Bolshevik collective. Named for Aleksei Grigorievich Stakhanov, who mined (that is, expended solar/caloric energy) far beyond his quota in a single day in 1935.

SUN

The building block of all life on the planet . . . sunlight (represented in rays) is focused and stored. Rays, we might say, are a way of representing the sun as fuel, before it has been captured and made into energy.[81]

The sun as fuel (or, rather, as a medium that can produce fuel) is not easy. Consider, for instance, *The Mysterious Island*. The colonists have collected wood (fuel), but they require ignition since the use of friction (a savage means of making fire) is not ideal. Returning from a mission away, Pencroft is astounded to see flames.

"But who lit it?" asked Pencroft.

"The sun!"

Cyrus Smith has managed to create a device for capturing solar fuel, using watch crystals filled with water and sealed together with clay, "a veritable lens, with which he had concentrated the solar rays onto a clump of very dry moss and so induced the process of combustion" (2001, 88).

FIGURE 6. Smiling sun with rays. *De re metallica.* 1556. Why does the artist depict a primitive smiling sun with iconic rays overlooking what is otherwise a didactic illustration of a mining process? Courtesy of the Columbia University Rare Book and Manuscript Library.

In writings around the sun, we often find no distinction between fuel and energy. Here, for instance, is the text from a brochure produced by Ford from the 1940 World's Fair: "Man can change the form of energy, but he can neither create nor destroy it. The sun pours energy upon the earth, where it is stored in many forms. By controlling the channels through which it flows, as it changes from one form to another, industry harnesses this energy for the benefit of man."

Victorian scientist William Thomson espoused the notion that all motion on earth derived its mechanical power from forces of the solar system or the sun's heat. But whence the heat of the sun? Thomson believed it derived from "friction in the atmosphere between his [!] surface and a vortex of vapours, fed externally by the evaporation of small planets, in a region of very high temperature round the sun, which they reach by graduate spiral paths, and falling in torrents of meteoric rain, down from the luminous atmosphere of intense resistance, to the sun's surface" (cited in Smith, 147). Around the same period, the **vril**-ya, a subterranean people in Edward Bulwer-Lytton's *The Coming Race,* manages to grow agricultural crops. We do not know how this is possible. We are told, simply, that the world below the earth's surface is quite temperate despite lack of sun.

In the 1950s Sunday comic strip, "Closer Than We Think," Arthur Radebaugh, a Chrysler executive and designer and illustrator, is quoted: "We know how to get electricity energy from sunlight by means of silicon converters, and if we are able to develop small, efficient energy storage cells[,] solar powered cars will be feasible."[82]

Today we might satisfy nearly all of "our" energy "needs" by directly harnessing the sun as a fuel. To do so, we would have to develop fully efficient photovoltaic cells and deploy them over .02 percent of the earth's surface (including oceans) (Prentiss, 33).[83] Why is this so unthinkable? Yes, such a new system of machines would transform traditional land and seascapes into futuristic cellularscapes. The traditional grid, partially hidden, partially visible, would also be disrupted with more localized inputs and new infrastructures. This machine would have certain costs associated with it and it might disrupt ecosystems. Such issues lie outside the immediate scope of this dictionary, and yet . . .

TALLOW

Fat from beef or mutton, tallow was used in candles before whale oil, before wax became commercialized and cheaper.

It can also be used for biodiesel, and since the fat has little other use (no longer being used to cook McDonald's french fries), it avoids the food vs. fuel dilemma.

Cyrus Smith informs the men that they will undertake a seal hunt. Why? To make tallow? "Fie, Pencroft! So we can make candle wax," replies Smith

(Verne 2001, 190). Since the colonists have already prepared lime and sulphuric acid, they can undertake a chemical reaction to produce oleic, margaric, and stearic acids. They make wicks from vegetal fibers. Their candles are not pretty: "they were neither bleached nor polished, nor were the wicks impregnated with boric acid to make them vitrify and disappear as they burned," but they do the job of lighting Granite House (191). The reader understands that the men choose to hunt seals because they have not yet done so, because they can.

The seal hunt in *The Mysterious Island* appears excessively violent, at least in the illustration (adorned with the caption: *Ces animaux vigouresement frappés*). A flock of birds crowds the sky. The men (with Neb in the forefront) wield sticks to slay the beached and helplessly immobile creatures. Why? Perhaps because Ferat, the illustrator, wishes to allow himself a certain dynamism that he has had to keep in check in much of his other work. Is this compositional aggression a kind of fuel?

THEOLOGY

We have already encountered the automaton as a figure that pretends to function on its own. Or rather, we prefer not to see who or what is pulling the levers. Now let us recall the famous opening of Walter Benjamin's essay "Theses on the Philosophy of History." Benjamin rips open the cover of a large box and demonstrates that the automaton or puppet called "historical materialism," which wins every game of chess, is actually, tragically, fueled by theology, "which today, as we know, is wizened and has to keep out of sight" (253).

THORIUM

Discovered by a Swede and named for Thor, the Norse god of **thunder**, thorium was once used to fuel gas mantles. It gives off radioactivity. Once this was understood to be dangerous, thorium gas mantles were dis-mantled.

THUNDER—SEE LIGHTNING

URANIUM

The primary fuel used in nuclear energy, uranium is very heavy and is found primarily in "our kind of place" (the United States, Australia, and Canada), as well as some developing countries, in recoverable veins or in dilute amounts in seawater.

FIGURE 7. Ces *animaux vigouresement frappés* ("Those vigorously beaten animals"), *The Mysterious Island*. Courtesy of the Columbia University Rare Book and Manuscript Library.

Uranium-238 is "fertile," that is, it can undergo the transmutation (into another element in a nuclear reaction) that Rutherford and Soddy dreamed of. It is precisely the fission of material that produces heat in reactors or produces the material for nuclear weapons. The element was probably first isolated in pitchblende and named after the planet Uranus.

You hop right over the door and into your Ford Nucleon, like Batman in his Bat Car (1958).[84] You fire up the on-board nuclear reactor, suspended between twin booms at the rear. And . . . why not . . . you drive out of the Bat Cave. From the television series, that is. Bronson Canyon, Hollywood Hills. You kick up dust and then hit the asphalt. But instead of heading down Canyon Drive to Hollywood Boulevard, you are in Gotham City, which is really New York, or maybe New New York.

You drive all the way from New York to Hollywood (or vice versa) and then back to Detroit without recharging.[85]

VEGETABLE OIL

. . . and animal fats are composed of three chains of fatty acids, bound by a glycerol molecule. These can be turned into biodiesel after the oil is filtered, water removed, and the free fatty acids neutralized by adding a base to the solution. The process is slow and messy, but you could do it in your backyard if you so desire.

VIS VIVA

Before the development of modern thermodynamics, *vis viva* or *Lebenskraft* (for German physicists) was considered a material element, a purposeful force of the universe, close to kinetic energy. The fuel of God's mechanics. Could it be destroyed, wasted? Was it subject to entropy? Physicist and brewer James Joule believed that only the creator can destroy (matter) and the universe is a perfect machine that suffers no loss. Then how to explain the (very alchemical) question of the putrefaction of matter? Could chemistry and physics alone explain the breakdown of organic matter as scientists observed it in animals?

VOLCANO (SEE ALSO **GEOTHERMAL**)

Almost as soon as Cyrus Smith gains consciousness after the balloon crash (his first utterance: "Island or continent?"), he informs his fellow colonists that they are on land formed by a volcano that is now extinct. The reader is not in suspense about this fact since she takes all that the engineer says on faith.

In *Theory of Literary Production*, Pierre Macherey notes that all literary works (*The Mysterious Island* in particular) deviate from what they appear to be. Verne's texts exhibit a remarkable coherence on the surface. Yet there is always some mystery. Macherey cites the writer Michel Butor to this effect:

> Indeed it is very important that the book itself include a secret. At the start the reader must not know how it will finish. There must be some change produced for me, I must know something that I did not know before, that I did not guess at, that others would not guess without having read it; and as might be expected this requirement finds its clearer expression in popular forms like the detective novel. (cited in Macherey, 36)

So in the case of *The Mysterious Island*: What is this mystery that causes a change in the reader? Is the denouement at the moment when Cyrus Smith calls together the men and repeats the secret that Nemo whispered in his ear? "My friends . . . Lincoln Island is not destined to last as long as the Earth itself. It is fated for destruction in the very near future. The cause of its annihilation lies within the island itself and it cannot be stopped!" (Verne 2001, 612). And lest the reader feel that Smith is being a drama queen, Verne continues with a bit of geological/contextual prose:

> It must be added that Cyrus Smith had in no way exaggerated this threat. Volcanoes are generally found in the vicinity of oceans or lakes, and it has sometimes been thought that a sudden flow of such waters into the volcano's chimney might very well extinguish them forthwith. But any attempt to carry out this plan would incur a strong risk of shattering a vast area of land, for the effect would be not unlike firing a bullet into a hot boiler. Rushing into an enclosed area whose temperature might be measured in the thousands of degrees, the water would be vaporized instantly, with such force that no wall of rock could ever withstand the pressure. (613)

Macherey labels this ending a "surprising transformation of Verne's initial project" and an admission of the failure of the energetic conquest of the island (Macherey, 48–49). So far, perhaps, so good: a sort of negation of the Anthropocene *avant la lettre*. And yet, if we read deep into Macherey's text, we come upon a strange remark, one based on a familiar theme, that is, a reference to the phrase *mobilis in mobili*, but here translated in a new way: "The literary failure of Jules Verne, the fragility of that enterprise which is not his alone, this is what forms the matter of his books: the demonstration of a fundamental *historical defect*, whose most simple historical expression calls itself the *class struggle*." This is the first time we come across this notion, at least stated so bluntly. Are we to understand that the real mystery of the island is not the peculiar volcanic geology but in fact another sort of rift related to labor and capital? "Today and tomorrow, *mobilis in mobili*—for the novel this involves two things at once: and yet they are not the same thing." The text contains conflicts: One of these is the encounter of the men and Nemo (a conflict completely created by Verne himself and not dictated by external or structural conditions). And at the same time, "Nature is not what one took it for: between that basic reality and human actions there is not only the mediation of labour and science, but also the entire screen of historical myths, factitiously constituted but none the less real" (238). Nemo is the father of a certain kind of bourgeois ideology, and the book itself does the work of demystifying the figure that he is. Verne aspires to represent a vast panorama of history, but "the book is not a simple reflection of the contradictions of its time, nor is it a deliberate description of the project of a social class at a given moment. . . . It is a perception rather than a knowledge in the true sense (a theoretical knowledge), rather than a mechanical (unconscious) reproduction of reality" (239). Again Macherey mentions a flaw or defect, but now it is not of the text but of Verne himself, and it relates "not to a *historical project* but to a *historical statute*." By this Macherey seems to mean not an ideological contradiction that is revealed to us through theoretical meditation, but rather: "In its way, so simple and curiously veiled, perspicacious and deceptive, the book finally shows us—though it may not be in the manner stated—what it proposed to enunciate: the history of its moment" (240).

The revelation of the secret of the volcano forces the colonists to work even harder to produce a "real" ship. That they can think, imagine a catastrophe that impels them to be ever more industrious, is in itself a moment of ideological coherence, undone by the volcano and yet intact in the rebirth of the new island.

Explosive volcanism—and not simply in the form of benign, constant, geothermal energy—harnessed as a fuel?

VRIL

I have already mentioned the fertile subterranean agriculture of Edward Bulwer-Lytton's Victorian sci-fi, *The Coming Race.* The human narrator falls down into a vast universe powered by a fuel called vril and populated by a race, the Vril-ya, named after said fuel. Vril, like **radium,** has many properties, some destructive, others life-giving. We never learn precisely what vril is. It remains untranslated. As the narrator explains (or better, mystifies), "I should call it electricity, except that it comprehends in its manifold branches other forces of nature, to which, in our scientific nomenclature, differing names are assigned such as magnetism, galvanism, &c" (Bulwer-Lytton, 26).[86] Vril can be inserted into motors, and it also works through the mind. For instance, it can be used to put the narrator into a trance during which he learns the language of his hosts. Vril lights the world and powers transport. The Vril-ya bathe in it a few times a year to increase vitality, but it can also be used as a highly destructive weapon as it also leads to peace and eliminates crime. Vril is contained in staffs or wands. The force of the wands is not equal for all: "but proportioned to the amount of certain vril properties in the wearer, in affinity, or rapport, with the purposes to be effected" (64).

The narrator goes to a museum that contains antiquated forms of transportation. The Vril-ya regard balloons and steam-powered vehicles with contempt, as pre-vril inventions of savages. Vril powers **automaton** servants, so there is no need for debased human labor.

WASTE

In the late seventeenth century, a British woman traveling from the eastern cathedral city of Peterborough to nearby Wansford—an area remote from water or coalfields—saw patties of cow dung drying. She remarked: "It is a very offensive fewell but the country people use little else in these parts" (Prentiss, 43).

For the purposes of this discussion, waste could be defined as anything that is not usable by humans, including **biomass.** In typical waste plants today, a certain number of items are removed, either because they can be recycled or they are not combustible or they are toxic. The greatest percentage of the waste is burned in incinerators in order to generate heat. We could stop our discussion here because what happens next does not concern fuel per se, so much as the economics and logistics of what is really a very old and inefficient technology (the Rankine system), exorbitant losses experienced by the condensing of the low-temperature, low-pressure steam at the discharge of the turbine. Waste as fuel is much discussed in political terms, since it tends to be localized, whereas electrical utilities tend to be regional or even national. And this is to say nothing of the toxicity of incineration . . .

Yet waste, paradoxically, also refers to the very dissipation or nonproductive use of *any* fuel. It is the duty of man, according to nineteenth-century Scottish scientists such as William Thomson, to "employ engines for the benefit of mankind and in aid of its commercial and moral 'progress.' Failure properly to direct and harness those gifts of energy was therefore only a waste, and in that sense a sin" (cited in Smith, 101). Thomson's peers were obsessed with waste. However, waste becomes something very different if we think about this term in a grand, planetary scale, where waste is abundant, the rule of law. Allan Stoekl warns us against thinking about waste as a fuel to allow us "the power to do work." "The very use of the term 'work,' albeit in this context a purely neutral, scientific term, nevertheless does tend to anthropomorphize energy: its power not only, say, moves or heats; it accomplishes something in a purposeful, human sense." Yet when we take a giant leap back (for mankind), we may see that "in speaking of the finitude of energy supplies, we are only speaking of the limits to the human, the fundamentally limited availability of ordered energy capable of doing 'work' for Man. We are speaking, in other words, of death" (200).

WATER

In some respects, water, like **air**, appears to be a free and pure form of fuel. Pure hope?

Hydropower functions by blocking a flow of water (in a dam, say) and converting the potential energy into kinetic energy as the flow is opened up. That kinetic energy is then transferred to a turbine, attached to a generator shaft that is rotated to produce electricity. By some calculations, water could fuel almost all of the energy needs of the United States now.

The Arrowhead Tear Drop concept car, 1936. Shaped like a teardrop and sponsored by Arrowhead Water. "Years ahead, the Arrowhead Tear Drop

Car reflects the trend of automotive designing. It is America's most modern car. And just as far ahead in design is the famous Arrowhead Hydro-cooler!"[87] Does the car have a water tank to be filled with water at water stations across the country?

Water (broken down into its component elements by electricity) will replace **coal** as a free and clean form of fuel. In the future, Cyrus Smith promises, water will be easily harnessed:

> Yes, my friends, I believe that water will one day be used as fuel, that the hydrogen and oxygen of which it is constituted be used, simultaneously or in isolation, to furnish an inexhaustible source of heat and light, more powerful than coal can ever be. One day, the holds of steamships and the tenders of locomotives will be filled with those two compressed gases, powering their engines with an incalculable calorific force. . . . I believe, then, that once the coal deposits have been exhausted, we will warm our homes and ourselves with water. Water is the coal of the future. (Verne 2001, 327)

Unfortunately, for the curious reader, the conversation is interrupted by the barking of the dog and the domesticated ape/servant.

The men of the colony of Lincoln Island develop a hydraulic elevator to take them to their apartments after apes sabotage their system of ladders. And like so many of the illustrations for Verne's work in feuilleton and book forms, the elevator is depicted as a modern, industrial contraption with welded joints, and not something pieced together with limited resources. The young boy reading this tale would be asked to take a leap of faith, to stitch together images of technology (and where did the apparently bereft colonists come upon so many tiny American flags to adorn their projects?) with a wholly disparate narrative. Such is the work assigned to the reader, but one imagines he is rewarded with a considerable pleasure. In the case of the elevator, then, water serves as a fuel, but one that is almost wholly eclipsed by the sophistication of the machine. In the difference between the perilous conditions laid out in the *robinsonnade* portions of the book and the elaborate and fine details of the images, we find, precisely, the elevation of a certain kind of didacticism that otherwise tries to hide itself in the prose of adventure.

Jack Harper (Tom Cruise) dreams of an old New York that he never knew in the 2013 film, *Oblivion*. It is 2077, as his voiceover instructs us. For some reason, a race of alien scavengers destroyed the moon (Why? Perhaps for **helium-3**? But then they would not have blasted it apart, one assumes), causing widespread geological change on earth. Humans, Jack explains, had to use nukes. Now the scavengers are gone, but humans have moved off-planet, first to a holding station called Tet and then to a terra-formed Titan, Saturn's moon.

We should know all of this if we were organically integrated into the logic of the film, without Cruise's Sciento-alogical milquetoast history lesson. We should find our way into this history without the sweeping panoramas of vestiges of familiar landscapes, now devastated, but the film takes a short cut: it doesn't waste our precious time, because we need to devote the bulk of it to fight scenes (between Cruise, who, inexplicably, wears an ab-enhancing flight suit on his daily sorties to the ravaged homeland and what we believe are Scavenger-controlled drones). Moreover, *Oblivion* wants us all to like it. Right and Left. Urban and Rural. Adults and Children. It cannot risk appearing partisan. Although we see a field and goalposts, and although Cruise reminisces about the QB and WR at the last Superbowl (2017), there is no mention of any particular team or city that might alienate fans of a rival, for instance.

Harper (actually a clone who thinks he is a man who underwent a required memory wipe) has been paired with a woman (also a clone, but perhaps a bit more aware or at least less rebellious and certainly more robotic, who, incidentally and inexplicably, wears business attire every day as she serves as his communications liaison seated before a screen and control panel). Vicka ("we are an effective team," they repeat to their boss, played by Melissa Leo, who turns out to be a holographic hologram), stays behind in their spaceage-mid(past)-century-modern pad while Jack goes off every day to protect the "hydrorigs," large floating containers in Soviet brutalist style that extract seawater, apparently to be subjected to fusion, in order to provide fuel for the human colony on Titan. (Is the water flown to Titan and subjected to the process there? Or does the energy itself transfer to Titan along some solar-system-wide grid? And why transport seawater all the way to Titan? If we've gotten far enough with our technology to move to Titan, surely we've found some water and/or an energy source there? And why is the "effective team" necessary? Why is Vicka even there, since Jack clearly has no sexual interest in her. Details, details. We will never know and we are supposed to forget—maybe we'd all benefit from periodic memory wipes. Never mind that in the end there is no Titan settlement and the water is actually being stolen by the

Tet itself, and never mind that we never establish why or how the alien vessel has need for water ("it's a resource"). None of this matters in the end because after one of the Jack clones blows apart the Tet (with **plutonium**), we cut to a scene of a lakeside cabin powered by small **wind** turbines (they look more like wind chimes) and solar panels that Jack has been keeping. Here, his real (human) wife is gardening and watching a toddler: the other humans on earth have found her and with them . . . Jack . . . well, not the same Jack who blew up the aliens (a long close-up on the number inscribed on his flight suit makes this, at least, crystal clear). But still, in some way, even this Jack must have some of the original Jack in him. And so life begins again . . . around a pond.

WHALE OIL

During a particularly industrious period (Smith is building a mill, Spilett and the boy, Harbert, have discovered tobacco and are secretly drying it to satisfy Pencroft's "very last desire on earth"), a whale is spotted off the island. With an exclamation that makes up for in boldness what it lacks in subtlety, the sailor cries: "What fine thing it would be to get our hands on that whale!" (Verne 2001, 313). On the other hand, the caption in the French text focuses on the whale as monster—almost as if its presence threatened the men. The enterprise of manumission seems impossible, so the men return to other tasks, but the sailor regrets, since he cannot do otherwise. "In every sailor lives a fisherman, and if the pleasure of fishing is directly proportional to the size of the animal, we can only imagine what a whaler feels in the presence of a whale!" For the sailor, the whale is a like a "forbidden object" for a child (314). Finally, though, the men do not harpoon it or struggle to capture it. The whale washes onto shore (steered perhaps by "the mystery" of the island?) and the men are able to congeal its blubber for fuel (not to mention the usefulness of the bones).

A good seven hundred pages into Charles Dickens's *Bleak House*, we find a chapter (number fifty-eight) titled "A Wintry Day and Night." Lady Dedlock is missing, suspected of nefarious deeds, but her husband asks that "good fires" be prepared so she will know she is expected:

> The day is now beginning to decline. The mist, and the sleet into which the snow has all resolved itself, are darker, and the blaze begins to tell more vividly

FIGURE 8. "Quel monster," s'ecria Nab ("What a monster!" cried Neb). Really? This whale does not seem to bother anyone. But as in Verne's *Robur*, where a whale apparently threatens the low-flying aerostat, it exists to be conquered. *The Mysterious Island*. Courtesy of the Columbia University Rare Book and Manuscript Library.

upon the room walls and furniture. The gloom augments: the bright gas springs up in the streets; and the pertinacious oil lamps which yet hold their ground there, with the source of life half frozen and half thawed, twinkle graspingly, like fiery fish out of water—as they are. (742)

The whale still inhabits the fuel, perhaps through its visual or olfactory impacts. Whale oil is a fuel that retains a sensory imprint of its animal origins.

WIND

In the Bayeux Tapestry, wind is shown in the sails of ships as a curving line. If wind had favored William the Conqueror as soon as his men and ships were ready, he would have set sail six weeks earlier. He would have encountered King Harold's force by sea and land, and would likely have lost, meaning there would be no Norman Conquest.

Leon Battista Alberti suggests that painters should include images of Zephyr or Auster blowing "in order to account for the seemingly unnatural postures of inanimate things, such as hair, leaves, clothes, when represented in a state of motion" (Stimilli, 77). In Agricola, we find a woodcut depicting a basic, iconic face placed in the sky, blowing gusts of winds—curly lines— from its puffed-up cheeks.

In the *De re metallica*, we come across an image of several hearths, including one dotted with **bismuth**. In the upper-left corner, "old man wind," an atavistic reminder of ancient personifications, serves an apparently decorative function. By the High Renaissance, such devices are already outmoded and wind is naturalized.

The wind carries the sailor where it wishes. It is out of his control. Before the railroads, the raw materials of the textile industry—wool and cotton— were carried by ships, powered by wind.

Jules Verne's colonists require a mill to process the wheat they have grown (from a single seed found in a pocket!). While they might well have used the second waterfall of the river they have called the Mercy, a "windmill was no more difficult to build than a water mill, and they knew their mill would never want for wind on this plateau, exposed as it was to the continual breezes of the open sea." In addition, while a water mill is limited by the direction of the course of the water, it is possible to make a rotating windmill whose face can be turned toward the breezes. Moreover, it does require great motive

Focus in quo uena excoquitur A. Focus in quo plumbi guttæ iacent B.
Forceps C. Alueus D. Ventus E.

FIGURE 9. Hearth with "old man wind" blowing, from *De re metallica.* Courtesy of the Columbia University Rare Book and Manuscript Library.

force: "Experiments have shown that a windmill performs with greatest efficiency when the number of revolutions the sails make in one minute is six times the number of feet traveled in one second by the wind. With a minimum breeze of twenty-four feet per second, the windmill's sails will turn sixteen times per minute, and no more than that is required" (Verne 2001, 382). So even here, on Lincoln Island, in the middle of nowhere with no industrial machines in sight, no need for machines to link up to other machines, no hope of reproduction of the species, it is still important for the men to establish a constant flow of energy.

FIGURE 10. Pencroft is delighted by his work. *The Mysterious Island*. Courtesy of the Columbia University Rare Book and Manuscript Library.

There are also aesthetic reasons behind the decision. A windmill makes a handsome landmark, Verne notes. (But for whom?) Pencroft, flanked by Neb, who is astounded by everything, takes great pleasure in his work. Aesthetic and energetic principles aside, it is clear that the purpose of the windmill in the novel is to provide readers with yet another engineering project to which they might aspire. It adds to the wealth of the already-diversified portfolio of solutions provided by Verne. (Note also the flag on top.)

Windmills sprung up in fields and towns of early modern Europe. Most of those still standing serve as emblems of the preindustrial past. Today, again, wind farms with large white turbines dot the landscape.[88]

Heidegger might caution us about placing faith in the wind turbines, not because they are expensive "now" (until an appropriate economy of scale can be achieved). Not because they are ugly or threaten bird migrations (although both of these may be true), but because of their relationship, precisely, to a certain idea of hope. In the past, Heidegger notes, man built a wind-powered mill that was geared into nature. When the wind blew, it turned, producing energy that could be used immediately (*unmittelbar*). Now the wind turbine as alternative-energy generator takes wind out of nature, stores it, and makes use of it in remote times and place. At least this is what he says, although for engineers in the sector the question of storage is certainly quite complex. Such storage or reserve is, for Heidegger, a form of rape in essence, not qualitatively different from a hydroelectric plant that steals water and makes the river no longer a river but a system of generation. He writes: "The revealing that holds sway throughout modern technology does not unfold into a bringing-forth in the sense of *poiesis*. The revealing that rules in modern technology is a challenging [*herausfordern*], which puts to nature the unreasonable demand that it supply energy that can be extracted and stored as such" (Heidegger 1977, 14). He will, of course, go on to speak of another kind of revealing—*aletheia*. Wind, for Heidegger, bears nothing in its essence that destines it for *Stellen*.

Readers of this dictionary may be familiar with Heidegger's essay, as well as his later writing on *Gelassenheit* or "letting go," "letting be," or "relaxing."[89] This is a difficult word to translate. For Heidegger, the modern definition of technology as an instrument—like the hydro-powered turbine—is technically "correct." Yet a primitive waterwheel placed in a stream to power a grain mill is not as challenging in the same way as, say, mining matters from the ground.

What happens when we put this thought to the colonists of Lincoln Island, for whom the mill is a "landmark" and a "project" for *ascesis*?

WOOD

In one version of this dictionary, wood—*fuel*—would be the only entry. Today the continent of Africa receives 90 percent of its power from wood. "Fuel," as we recall, is etymologically related to the *focalia*, the ancient law of the right to gather wood. In the early Middle Ages, wood courts regulated the rights to access in the communal forest. Each member of a community was entitled to wood for construction, tools, and firewood. Various restrictions were imposed on activities such as clear cutting or the removal of certain valuable trees. It appears almost as if John Evelyn cannot help slipping into the fuelish when he writes of those who would violate laws protecting trees: "I dare not suggest the encouragement of yet farther restraint, that even proprietors themselves should not presume to make havock of some of their own woods, to feed their prodigality, and heap fuel to their vices" (Evelyn, 2:146). Those woodlands that belonged to the *desmesne* of the crown were withdrawn from general use. Their use was out—"*foris*," which is the origin of "forest" (Sieferle, 70). Indeed, the perceived scarcity of woodlands often led to increased regulation. The *focalia*, the origin of fuel, we might say, is intimately tied to the development of the absolutist state and sovereign law.

Silas Marner has moved from Lantern Yard, a rapidly modernizing town, in a valley, to Ravaloe, a woody one—a fuelish town still immersed in the past. George Eliot writes: "In the early ages of the world, we know, it was believed that each territory was inhabited and ruled by its own divinities, so that a man could cross the bordering heights and be out of the reach of his native gods, whose presence was confined to the streams and the groves and the hills among which he had lived from his birth" (15). In contrast, Ravaloe is a "low, wooded region, where he felt hidden even from the heavens by the screening trees and hedgerows" (14). Moreover, "Ravaloe lay low among the bushy trees and the rutted lanes, aloof from the currents of industrial energy and Puritan earnestness . . . fears were like the rounds of beef and the barrels of ale—they were on a large scale, and lasted a good while, especially in the winter-time" (23).

Like the men of Lincoln Island who pass the time without boredom, in Ravaloe, Silas Marner keeps a warm hearth going in his cottage and his door open. Dunstan, a local man without principles, down on his luck, wanders in the cold and approaches the cottage with the idea of stealing money from Silas. Silas is away, perhaps "to fetch in fuel," Dunstan imagines (Eliot, 39).

Shortly afterward, the hearth again plays a crucial role: A child, her mother fallen into a stupor in the snow, sees a bright light in the distance. The little one, "rising on its legs, toddled through the snow, the old grimy shawl in which it was wrapped trailing behind it, and the queer little bonnet dangling at its back—toddled on to the open door of Silas Marner's cottage, and right up to the warm hearth, where there was a bright fire of logs and sticks" (115).

Marx defends the right of the poor to forage for branches fallen from trees on the forest floor against the edicts granting property owners the right to assess value and punishment for theft. Indeed, it is the context of his piece on wood that Marx writes what will become a signature question, at times misunderstood, wrenched from its context like green branches that are wrenched from young trees: "If every violation of property without differentiation or further definition is theft, would not private property be theft?"

Before the expansion of the textile industry and the introduction of steel, weaving machines—the water-frame, spinning jenny, the mule—were made from wood.

In Bangkok, a bo tree (sacred to Buddha) falls in a canal. "And despite the shortage of wood, no one will harvest it either. It would be unlucky" (Bacigalupi, 79).

As the novel reaches its bleak conclusion, the Windup Girl meets a man and woman who have also survived mass chaos and the bursting of the levees. Together they "light a fire of kindled furniture on her apartment's [high floor] balcony and roast . . . fish" (357). The fire is intimate, domestic, survivalist, and warm. At first Emiko is diffident toward the man, but the hearth, the fuel, draws them together. Emiko laments that she can never reproduce. The engineer replies, "I cannot change the mechanics of what you already are. Your ovaries are non-existent. You cannot be made fertile any more than the pores of your skin supplemented." When Emiko looks crestfallen, he continues: "Don't look so glum! I was never so enamoured of a woman's eggs as a source of genetic material anyway. Any strand of your hair would do. You cannot be changed but your children—in genetic terms, if not physical ones—they can be made fertile, a part of the natural world" (359). A city whose population has been devastated (or cleansed) by massive floods. A new beginning; an offer by the (male) scientist to help the obedient courtesan-automaton to have a child (actually, to do it for her—through his mastery) so that a race of New and perhaps redeemed people will come. This is a kind of hope—sparked by fuel—but not an authentic one, a hope in bad faith.

Yet wood might lead us to a more genuine hope. . . . It is useful to remind ourselves here that the Greeks used the same term—*hyle*—for forest trees and matter itself. Fuel, wood, and material(ism) form a knot: "So it was not for nothing that the very *name* (which the Greeks generally apply'd to timber) *hyle*, by *synecdoche*, was taken always *pro materia;* since we hardly find anything in *Nature* more universally useful; or, in comparison with it, deserving the name of material; it being, in truth, as the mother parent and (metaphorically) the passive principle ready for the form," as Evelyn writes (2:119).

ZOLINE

Carl Fischer of Pittsburgh wrote to Ford to announce he had invented a secret powder that could be mixed with water to drive automobiles at a cost of three cents per gallon. Apparently Fischer stripped naked and entered a cell with only a gallon of seawater. When he emerged (presumably he held the zoline in his mouth), he poured the water into a marine engine. It ran like a charm. Although Ford's secretary, Dailey, carried on a correspondence with Fischer, the industrialist himself was apparently too busy or disinterested to investigate further (Wik, 78).

ZYKLON B

A fuel for killing—bugs and people. A gas, consisting of hydrogen cyanide, a stabilizer, and a warning odorant (eliminated in the concentration camps). Bayer, makers of aspirin, and today, of pesticides for first-world pets (brand name Advantix), was one of the primary manufacturers. A fast, efficient fuel that does not have immediately recognizable use for insertion into an energy-generating machine, but still one that moves bodies to change form. A fuel that should never again be used . . .

If (a) fuel could speak, would it ask to remain in the can, on the shelf, in the subsurface, let be, in its essence? Would it express a desire to remain unused, in *potentia*? And then, can a fuel possess the capacity for impotentiality?

A human possesses an ability, let us say, to manufacture or use a fuel. One could well decide not to do so. Such a decision concerns knowledge and power in the realm of the human.

Consider (or do not—you, reader, have the choice to not do so) a treaty produced in Ecuador to not take oil from the ground.[90] To be sure, this treaty reflects several different, possibly competing strategies or goals. One is simply a global effort to preserve extreme biodiversity in the Yasuní National Park (which sits on top of a large oil reserve). The reserve is subjected to a surface vs. subsurface debate at stake as in various parts of the globe around issues of resources. Another aspect of the treaty is to use funds collected from crowdsourcing and lobbying of nations and NGOs to help alleviate poverty of indigenous people in the region. A third is that some of the donations have been promised toward the development of alternative or "future" fuels, thus helping reduce Ecuador's dependency on oil. But most significant, the Ecuadorian government is asking for funds to NOT use the oil, in part, as a way of keeping four hundred million tons of carbon dioxide from entering the atmosphere. How can critical theory help us think this international effort? Does it represent a mode of (hopeful) potentiality?

Readers of this dictionary may be familiar with debates around the Keystone XL Pipeline. Environmentalists have made of it a symbol—and its impotentiality a symbolic gesture. Build the pipeline, do not build the pipeline. Build it and you will create jobs, which may be only temporary, or permanent, minimal or numerous. You will disturb landscapes sacred to Native Peoples, you will improve the lot of small farmers struggling against drought and factory farms, who can now toil less and retire, and you will risk terrible leaks that will seep into the soil and pollute aquifers and disrupt local ecosystems. You will create a scar on the land that will cross a border and crisscross through the plains down to the sea where the contents of the pipeline—assuming it has not spilled or clogged up—will be refined and repackaged and separated out and loaded on ships that barrel through channels or enter other pipelines. Or the tar sands will enter into other pipelines, already in place, crossing the border and crisscrossing the landscape or will be loaded into railcars and onto ships. It makes so little difference, really, to the molasses-like ooze that is willing to be pushed or packaged or separated out, but it—oil and sand together—is also willing to remain in the ground. In fact, that is its default position.

For Agamben, *dynamis*, far from being a quality of the physical world, has to do with Being. He glosses Aristotle: "If a potentiality to not-be originally belongs to all potentiality, then there is truly potentiality only where the potentiality to not-be does not lag behind actuality but passes fully into it as such." Put another way, *dynamis* preserves itself as such in *energeia*. "What is truly potential is thus what has exhausted all its impotentiality in bringing it wholly into the act as such" (183). Rather, as we might read Agamben, contrary to the idea that potentiality is annulled in actuality (or, we might say, that fuels are consumed in energy), in his reading of Aristotle we find a potentiality that conserves itself precisely in actuality. For Agamben, the alpha-privative in front of *dynamia* does not mean impotence or impossibility, but rather the (positive) capacity "not to" (do something).[91]

Jacques Derrida complicates any neat distinction between these two key words in the current age, the age of globalization, real-time communications, and the media. He speaks, perhaps somewhat uncomfortably, of a new era, one in which we can no longer authentically distinguish between the virtual and the real, "in the way one used to distinguish between power and act, *dynamis* and *energeia*, the potentiality of matter and the defining form of a *telos*, and hence of progress, etc. This virtuality is now inscribed in the very structure of the event produced; it affects both the time and the space of the image, the discourse, the 'information'—in short, everything that connects [rapporte] us to the said actuality, to the implacable reality of its supposed present" (Derrida 2001, 89). While Derrida does not, here, mention climate change or the Anthropocene (he was writing before these terms were in wide circulation in the public sphere), it is not far to carry his words to our present, devastating context. His words, though, resonate with bits of digital code and not so immediately for the question of a matter that could or could not be inserted into a system. In other terms, the wires and circuits and semiconductors that transmit pulses at speeds unthinkable for a recent past are not machines with boundaries distinct from the inputs, and indeed, at some point, the machine requires a fuel to make it work, and that fuel, today, is likely to be coal or water or uranium, whose traces appear nowhere in the bits themselves.

Perhaps Agamben's Heideggerian thought holds a kernel of hope. To be sure, as Antonio Negri might argue, Agamben would not lead us to any kind of action—neither a revolution that would finally "solve" carbon with an alternative to fossil fuels nor a slow but steady de-carbonization. Certainly, reading Agamben does not elicit a hope that we will discover a form of carbon capture and storage that—for whatever political or technological reasons—will garner enough support, perhaps through a global carbon pricing scheme, to stabilize emissions (back down) at the target of 350 ppm desired by climate scientists. Or higher, as we lower our expectations. Not a hope for a green revolution, an occupation of energy producers. Finally, we are not here getting at a form of individual behavior in conservation (of energy). Conservation might have an impact on global greenhouse gas emissions as a part of a larger strategy of diverse approaches. Consider, for instance, Stephen Pacala and Robert Socolow's wedges, a scheme that amortizes and diversifies different forms of sacrifice over time, but relies on actual present technologies. So compelling! So why aren't we following their scheme . . . today? Or rather, yesterday, since today may be too late.

Ultimately, conservation is merely the weak-minded reform ensuring expenditure, analogous, I think, to the Factory Acts of Marx's time ensuring working conditions (with a slight reduction of hours for some individuals). Conservation, far from oppositional, seems like part of the carbon-capital nexus just as factory legislation is as much a part of modern industry as "cotton yarn, self-actors, and the electric telegraph" (Antonio, 199).

This dictionary opens the idea that fuels can give (us) a gift. Let me be clear: by gift, I mean to invoke a kind of exchange that should require a return gift. What can we give back to fuels? Here is Agamben: "Potentiality, so to speak, survives actuality and, in this way, gives itself to itself." If, in the age of carbon intensity, language—figurative, analogical, magical—does not convey to and has no power over that which is not Being, then no matter how much we grant hope to "future fuels," they cannot return hope, cannot return a gift, only wait for us to (not) use them. And yet, if we were able to engage with them in more imaginative, perverse, creative ways . . . ? If epic is a departure from fuel and a return, then fuel figures in epic as the most conservative, closed figure. But what if we focus on what happens along the way? We will find fissures in the edifice of a self that can only be thought through reading the poetics of the text. What if we read this voyage from fuel to disorder and back as an opening to another way of thinking, through language . . . without closure, without end, but as potentiality even if—or supposing, rather, that—this disruption would mean a new way of living with intermittent power. Why is this so unthinkable? Must we in the first world adhere to an absolute equation of on-demand energy with hope? And then, why hope? Hope is an ambivalent and excessive term that should resist being molded to conform to energy policy, regulatory schemes (however "progressive"), or subsidies for the promotion of certain forms of fuel over others. Actions such as carbon-trading schemes, reducing pollution from existing power plants, or raising emissions standards for cars (if these are even feasible goals to attain in the current climate) are in no way adequate to hope, which can only begin to be reached by a gesture, precisely, of rhetorical violence. Hope, whatever else it may mean, certainly requires a leap of imagination, a way of thinking otherwise about the future. Perhaps, as anthropologist Hirokazu Miyazaki has suggested, hope might even constitute something like a method, as inspired by Walter Benjamin's notion of messianic history seizing "hold of a memory as it flashes up at a moment of danger." Benjamin writes: "Only that historian will have the gift of fanning the spark of hope in the past who is firmly convinced that *even the dead* will be not safe from the enemy if he wins" (247). The materialist historian must be ever vigilant against the enemy. He does not operate as a fuel, but more like a catalyst to keep the flame burning once it has caught on a combustible material. The spark of hope is ignited by "radical temporal orientation" (Miyazaki, 23). For Miyazaki, what is crucial is that the courageous historian "carves out a space for hope by changing the character of the direction of historical knowledge," and he applies this notion to his fieldwork as a method; in "the sparks of hope that have flown up from my encounter with the hope of [the Suvavou people] . . . mostly products of incongruities between the temporal direction of my own anthropological intervention and that of the Suvavou people's hope" (24). Such a method must resist an "immediate demand of hope for synchronicity" but must instead find hope precisely in nonsynchronicity. Long before the label Anthropocene was applied to the age of industrialization and fossil-fuel use, Benjamin understood that nonsynchronicity was not simply based on human actions but was also a disjunction between human time and geological time when he wrote: "The present, which, as a model of Messianic time, comprises the entire history of mankind in an enormous abridgment, coincides exactly with the stature which the history of mankind had in the universe" (Benjamin, 263). Fusion, fission, combustion, animals flapping their wings—these represent different temporal modes, different machines, connected by uneven grids. Perhaps a form of hope in this time of radical disruption.

In the very recent past (a past that has been rendered more distant than it might have been due to the speeded-up temporality, the acceleration implied by the Anthropocene), some held out hope that hydrogen would exist not as a commodity but a shared resource, an energetic commons (Rifkin, 216). It would be so not because of some cataclysmic shift (sci-fi dystopias—*The Windup Girl*, among other works, teaches us that "after climate change" bribes and scams will dominate public life), but because it will be so easy to generate and distribute without meters, without infrastructure and grids. Reading Macherey reading Verne calls to mind the stubborn quality of narratives. Once the writer/reader gets his talons latched onto a story, a process of petrification (a process allied with the formation of fossil fuels, but after unfathomable units of time) sets in. Talons and narrative form a bloc. They are repeated so often that it seems impossible to dislodge them or undo their bond. Perhaps fuels can begin to do so.

NOTES

1. In a recent study, a group of British scientists have quantified (in both monetary and purely physical terms) how much oil and coal already discovered would have to remain in the ground in order to avoid "catastrophic warming." The study is based on the assumption that cheaper reserves would be tapped first, before "tough oil." The authors are able to paint a very real scenario about losses already to be suffered by large energy companies and nation-states: they name names. Now clearly this study, which imagines *not using* over 90% of U.S. and Australian coal as well as almost all of Canada's oil sands ($20 trillion) is a concretization of potentiality, yet potentiality also exceeds it. Leaving fuel in the ground, then, also means a special relationship to it that is not just nonconsumerist. "Potentiality," with all of its philosophical ambiguity, seems well suited to recognize the complexity of the current moment.

2. Forty-three of the five hundred companies in the Standard and Poor's S&P 500 index are "energy companies," which makes strategies for divestment by universities or other corporate bodies especially complicated.

3. "A boy . . . A Water wheel . . . and a dream!" Advertisement. Duke University Library, J. Walter Thompson Collection, FM Box 1, FMC 1944 General, Domestic Ad. 1944–46. The ad was widely printed in farm publications.

4. Antonio Negri's key notion of "constituent power" is a combination of *potere* (power) and *potenza* (strength). As Maurizia Boscaglia expands in a translator's note to *Insurgencies* (*Il potere costituente*), for Negri *potere* refers to the existing power of the state and institutions. *Potenza* is a "radically democratic force that resides in the desire of the multitude and is aimed at revolutionizing the status quo through social and political change. Strength is at the core of the concept of constituent power itself as the force that produces (but cannot be contained within) power and its institutions; constituent power is fueled [!] by strength" (translator's note 3, in Negri, 336). Negri's work around biopolitics certainly appears addressed to human capacity for productive action and cannot be immediately translated into the realm of matter.

5. See Stoekl's *Bataille's Peak* for a brilliant discussion of postsustainable futurity: a future that might offer an alternative to austerity and sacrifice through excess energy; a solar energy so immense that it cannot be metered, monetized, or contained. A Bataillian future might not sit well with those policymakers who demand practical solutions, but his imagination seems at least to approach the unfathomability and "exorbitant" nature of climate change.

6. Dominic Boyer and Imre Szeman are two important proponents of an "energy humanities," a field that they note as being recognized by "the sciences, by government, indeed by industry." See their coauthored piece, "The Rise of Energy Humanities: Breaking the Impasse," February 12, 2014, www.universityaffairs.ca/opinion/in -my-opinion/the-rise-of-energy-humanities.

7. In "Murmurations: 'Climate Change' and the Defacement of Theory," his introduction to the edited collection titled *Telemorphosis*, Tom Cohen addresses the sovereign debt crisis in relation to attempts to "save" the humanities and to questions of climate change itself.

8. The text of the ad reads: "Go Green without Going Slow." "Driving a fuel-efficient* car shouldn't mean sacrificing performance. It's this belief that drove us to develop the TDI Clean Diesel engine and a turbocharged hybrid. Both help maximize driving dynamics while helping minimize environmental impact. And we've expanded that thinking beyond cars to everything we do; from the first LEED Platinum-certified automotive plant to working with the Surfrider Foundation to protect our oceans. It's called thinking blue and it's how we're thinking beyond green. *That's the power of German Engineering*" (*see www.fueleconomy.gov for mileage estimates). While in the public sphere, cars and transportation, individual consumption, and choices about fuel are dominant in the broad scheme of things, all forms of transportation (including planes, trains, automobiles, boats, and so on) account for less than 15 percent of global greenhouse emissions. In the United States the percentage is higher. Still, the car must be put in proper perspective. For Allan Stoekl, the car is the significant metonymy for a larger question of expenditure linked to fossil fuels, against which he will posit a bodily excess rather than a mere conservation of energy. To be sure, switching to electric cars, even if fossil fueled, would be more efficient than gasoline. Electric cars have smaller engines and no transmission, which makes them lighter, for instance. But long before the electric car was invented, killed, and revived, in Jules Verne's *Robur*, a group of Philadelphia balloon enthusiasts take their "Go-Ahead" for a spin when they encounter another flying object. At first, the men on the ground wonder if it is the flapping of bird wings. But then, "a suspicion communicated itself electrically [!] to the brains of all on the clearing" (Verne 1911, 139). It is the Albatross, the superior and powerful electric aerostat! For a succinct summary of the "clean diesel" swindle, see Taras Grescoe, "The Dirty Truth about 'Clean Diesel,'" *New York Times*, January 3, 2016, Sunday Review section, 7.

9. Ford Gyron Brochure, c. 1954. Collection of the author.

10. Zack Kantor, "Autonomous Cars Will Destroy Millions of Jobs and Reshape the U.S. economy by 2025," blog, http://zackkanter.com/2015/01/23/how-ubers-auto nomous-cars-will-destroy-10-million-jobs-by-2025/.

11. Two recent works resonate with the form of this dictionary. Hugh Raffles's *Insectopedia* includes (alphabetized from "Air" to "Zen and the Art of Zzz's") thought-pieces related to insects in the broadest sense. His encyclopedic form is, he might say, organic. By this he means that "the heroic protozoa created the planet's first encyclopedia by turning themselves into mitochondria and chloroplasts within other cells, which in turn formed alliances that grew into yet other beings, which joined up with yet others to make invisible cities, worlds within worlds. . . . Sometime after that time but still long before our time, there were the insects" (Raffles, 3). And Reza Negarestani's *Cyclonopedia* is a work—difficult to classify, part philosophy, part hallucination—about empire, narrative, and oil. *Cyclonopedia* (totalizing or comple-tist, but not alphabetical in form) makes for very compelling reading. It places oil above all other fuels (and perhaps rightly so), whereas this dictionary has different aims. To some extent, as should be obvious, I also began this project with Raymond Williams's *Keywords* in mind. While that work was meant to correct a gap between various cultures and classes in a period of rapid social change, as my work evolved it became clearer that what was at stake was not so much different uses of common terms so much as a broad mystification and disordering of material substances.

12. This is a key point in Angus Fletcher's *Allegory: Theory of a Symbolic Mode.*

13. This was the project of an excellent series of short essays in *PMLA* edited by Patricia Yaeger. As she notes, the essays were meant to peruse "the relation between energy resources and literature" (305). Some of the "energy resources" in this series could be classified as fuels in the sense of this dictionary (wood, tallow, coal, oil). Understandably, the authors of these essays do not necessarily need to separate fuel from energy to make their points. In fact, in his essay, focused on oil, Imre Szeman advocates for the importance of literature, of narrative, in relation to energy more generally. Literature is crucial in order to "shake us out of our faith in surplus (there will always be more; things will always be better), not by indulging in the pleasures of end times or fantasies of overcoming energy limits but by tracing the brutal con-sequences of a future of slow decline, of less energy for most and no energy for some—a future that might well have less literature and so fewer resources for manag-ing the consequences of our current fiction" (Szeman, in Yaeger, 325). Here it should be clear that Szeman is talking about both a form of fuel (oil) and a total system of which oil is an essential component.

14. Verne does not mention oil, but that is understandable given that the first well was drilled—in Pennsylvania—only about a decade before he published *The Mysterious Island* and oil was not widely used as a fuel until later in the century. Of the various critics who have written of Verne, Michel Foucault notes that some of the

scientific discourse reads like "immigrant" writing, inserted rather awkwardly into his narratives. For this, see Unwin, 193. Thanks to Anne O'Neil-Henry for this reference.

15. This is one translation offered by Macherey (162), but in addition, at other points he offers, "man is in things" (181); progress or journey (188); "masters of the world, masters of tomorrow" (209); science as "the essential instrument of a transformation of nature" (229); and "today and tomorrow." Elaborating on this last "translation," he notes: "The literary failure of Jules Verne, the fragility of that enterprise which is not his alone, this is what forms the matter of his books: the demonstration of a fundamental *historical defect*, whose most simple historical expression calls itself the *class struggle*. Accordingly it is no coincidence that all of Verne's images of reconciliation open out on to a description of a conflict. Today and tomorrow, *mobilis in mobili*—for the novel this involves two things at once: and yet they are not the same thing. An attentive reading will prove this" (238). In the general scholarship on Verne, the "correct" meaning of the motto is not immediately clear. In fact, in the Hachette edition of *Twenty-Thousand Leagues under the Sea*, the phrase is "mobilis in mobili" (plural), whereas Hetzel later corrected this to read "Mobilis in mobile." However, in *The Mysterious Island*, Hetzel returns to *mobilis*. The fact that there is no easy translation and that Verne himself moves from one version to another suggests that the inscription (that is, something fixed permanently as Nemo's motto) is itself mobile and fluid. The original 1875 edition of the work included illustrations by Jules-Descartes Ferat. See Evans, 118–19, for the status of images in Verne's *Voyages extraordinaires*.

16. According to Vernant, when Hestia agreed to remain unmarried, Zeus promised her that she would reign over the center of the house (*mesô oikô.*) The hearth is a symbol of centeredness, the *omphalos*, immutability. Human space is organized around it, and it is the men who go out, in centrifugal motions, from this center point.

17. I discuss Silas Marner and alchemy in detail in my *Alchemical Mercury*.

18. It seems especially interesting to consider, in the broader context of mining, fuels, and biopolitics, the movement in Latin America today known as Neo-extractivism. This movement, a reaction against the neoliberal privatization of natural resources of the 1980s, itself a movement in relation to the quasi-feudal conditions of labor that dominated in the colonial periods, is deployed with nationalist and progressive rhetoric. This is not the place to elaborate on the multiple geological-political stratifications involved in Neo-extractivism or the multiple ways it channels revenue streams to national governments through the same unsustainable and devastating relations to local ecosystems as ever; or the absence of a direct discussion around climate change; not to mention the fact that in many countries, the reclaiming of revenues bolsters and is bolstered by WTO monetary policies, for instance. In his dizzying historical/political account of fossil fuels, *Carbon Democracy*, Timothy Mitchell counters the idea that miners were able to collectivize easily because of their relative autonomy from management in the subsurface. "Strikes became effective," he writes, not because of a new, shared consciousness or the miners' isolation, "but on the contrary because of the flows of carbon that connected chambers beneath the

ground to every factory, office, home or means of transportation that depended on steam or electric power" (21).

19. In this regard, air is different from **wind**, or better, winds, which are personified by different gods in the ancient world depending on their strength and directionality. See figure 9 for the embodiment of the winds in early modern visual culture.

20. The Google self-driving car in development also resembles the Air pod car.

21. A recent dossier on MDI, Nègre's Luxembourg-based company, reveals that consumers are highly skeptical about claims regarding zero carbon emissions. See "Commenti e risposte," *La repubblica*, June 7, 2012, on the web. See also http://motori .ilmessaggero.it/motori/arriva_airpod_l_amp_39_auto_ad_aria_compressa_prodotta _in_sardegna_in_vendita_in_autunno/notizie/771967.shtml.

22. The term "prime mover" is a complex one. While in a pre- or early modern context it might have the value of a divine movement, a *deus* (*ex macchina*) or even **perpetual motion**, in the context of modern industry it can refer to a central or paradigmatic technology (for instance, the internal-combustion engine).

23. Krishna Ramanujan, "Beating Bird Wings Generate Electricity for Data Collector," *Cornell Chronicle*, March 10, 2015, http://www.news.cornell.edu/stories/2015 /03/beating-bird-wings-generate-electricity-data-collector.

24. Benson Ford Research Center, Dearborn, Michigan (hereafter FORD), 0069: Henry and Son Laboratories Series 1913–19, Box 1, Alcohol (as a gasoline substitute), correspondence.

25. The 1842 *Encyclopedia Britannica* defines energy as "a term of Greek origin, signifying the power, virtue or efficacy of a thing. It is also used figuratively, to denote emphasis of speech." See Crease, 418. In the first half of the eighteenth century, philosophers begin a broader discussion of how to understand, measure, and characterize terms like "activity," "work," and "force." By the early nineteenth century, scientists begin to group together phenomena such as falling water, expanding air, heat, and electricity. By 1899, Crease notes, the entry on "energy" in the *Britannica* is six pages long. This begs the question: Is it legitimate to speak of energy prior to 1800? Bruno Latour might respond in the negative, if by "energy" we mean "an abstractly defined concept of a quantity conserved in closed systems and related in precise quantitative ways to other concepts like power and work, no; if this means this is how we currently understand the world and its phenomena, yes" (Crease, 421). Following such logic, then, it would seem that fuel is much older than energy.

26. In a much-cited passage that is crucial for Agamben's thought, Aristotle wrote: "In its originary structure, dynamis, potentiality, maintains itself in relation to its own privation, its own sterēsis, its own non-being." And then: "A thing is said to be potential if, when the act of which it is said to be potential is realized, there will be nothing impotential" (cited in Agamben, 182). Agamben interprets this enigmatic passage to mean that the property of adynamia belongs to all dynamis. For Agamben, here, "beings" would seem to refer to humans. He makes a distinction with other life forms when he clarifies: "Other living beings are capable only of their specific

potentiality; they can only do this or that. But human beings are the animals who are capable of their own impotentiality. The greatness of human potentiality is measured by the abyss of human impotentiality" (ibid.).

27. When he began work on *Twenty-Thousand Leagues under the Sea*, Verne apparently crafted Nemo as an understandably misanthropic Pole whose family was killed by Russians. His editor, Jules Hetzel, insisted he change the character's situation because the editor distributed his books in Russia. In *The Mysterious Island*, then, Verne was able to insert a critique of Great Britain as an imperialist power. Chesneaux outlines the vicissitudes of Verne's political positions.

28. Agricola's text, more than 500 pages, includes 280 woodcuts.

29. For the relation of film to climate change, see my "Antonioni, Cinematic Poet of Climate Change," in *Antonioni, Centenary Essays,* edited by John David Rhodes and Laura Rascaroli (London: Palgrave Macmillan, 2011).

30. The quote from Ricardo and from Sismondi's *Nouveaux principes d'économie politique* . . . (Paris, 1819), T. II, 331, is an excerpt from Eugéne Buret's book *De la misère des classes laborieuses en Angleterre et en France* . . . (Paris, 1840), T. I, 6–7n. I might have included "slaves" as one of the fuels for this dictionary, but then I would have had to extend to all or every other human form. See Nikoforuk (2014) for the question of "energy slaves."

31. FORD 0069, Henry Ford and Son Laboratories Series 1913–19, Box 1, Alcohol (as a gasoline substitute), correspondence.

32. See Prentiss (45) for the hidden costs of ethanol. Ethanol is much less efficient than gas. One gallon of E85 gets only 70 percent of its energy from gas, so, in essence, consumers pay extra. Moreover, burning ethanol releases approximately twice the ozone as an equivalent quantity of gasoline. It also has a damaging effect on cars and can lead to increased grain prices, of course.

33. Thanks to Seth Michelson, who first brought the alchemical nature of Khosla's projects to my attention. In Khosla's process, methane from cow manure is used as fuel. Some corn must be transported from nearby farms, but overall, for every BTU of energy used to run the plant, five are produced. The typical ration for a corn ethanol plant is 1:3.

34. Peter Lamborn Wilson, Christopher Banford, and Kevin Townley, in *Green Hermeticism: Alchemy and Ecology* (Great Barrington, Mass.: Lindisfarne Books, 2007), attempt to link eco-thought with alchemy. However, they understand alchemy as a rather one-dimensional hermetic practice rather than a heterogeneous field in which theory and practice are ambivalently intertwined, as I discuss in *Alchemical Mercury.*

35. In *Alchemical Mercury*, chap. 4, I expand on the relation of "Rumplestiltskin" and alchemy.

36. The term "veer" here is important. See my essay, "The Risks of Sustainability," in *Criticism, Crisis, and Contemporary Narrative: Textual Horizons in an Age of Global Risk*, edited by Paul Crosthwaite (London: Palgrave Macmillan, 2011): 62–80.

37. As Verne notes, sulfuric acid, a crucial matter for modern industrial society, is made in large factories. The process he narrates, without machined tools and acid-proof containers, is one practiced in Bohemia for the production of Nordhausen acid (Verne 2001, 165).

38. FORD, "DESCRIPTION of the Dankwardt Process for producing GASOLINE from Crude Oil—Patent No. 1141529," Acc. 69, Box 1, "File: Distillation of Crude Oil. The patent was filed March 4, 1914, and was granted on June 1, 1915.

39. In *Objectivity* (39), Daston and Galison write that an epistemological break in the history of science occurs when the scientist is no longer supposed to have a pure soul.

40. Marinetti, 134. Translation mine.

41. Verne and Wells read each other. Verne estimated that his own work was much more linked to contemporary experimental science, while his contemporary projected images far into the imagined future. For his part, Wells may have believed that he was a more literary writer. Zola also disdained the unliterary qualities of Verne. See Unwin, 9–10.

42. At the Alberta tar sands, various actors repeat the fact that the Hudson's Bay Company used bitumen to repair roofs and the native people used it to seal their canoes. Of course this history does not compensate for the devastation of the land where the Metis and First Nations people now live. Nikoforuk (2010) offers a detailed discussion of the tar sands.

43. Adam McLean, commentary to *The Chemical Wedding of Christian Rosenkreutz*, 142.

44. Agricola may have been the first writer to use the term "petroleum." In his *de natura fossilium*, he mentions "the oil of bitumen . . . now called petroleum." See Agricola (580n).

45. However, somewhere around 300 million years ago, something happened to the fungi. The first white rot fungi appeared on the scene, able to digest the lignin. This meant much less coal—because the entire biomass was digested, and released as carbon dioxide. Genome sequencing has helped to clarify this.

46. See Hack for a detailed description of early mining.

47. Zola also drew on other strikes from the region. He set the novel earlier, in the 1860s. In part, this shift allows him to point to some positive changes that did in fact take place in the intervening period. For instance, by 1874 a law had been passed that made it illegal to employ women or children under the age of twelve in the mining pit. The Paris Commune took place in 1871. Trade unions were made legal in 1884. Still, the technologies of coal mining that he described did not change materially in the period between his visit and the novel's publication.

48. The bibliography on Zola's "Rougon-Macquart series" is vast. Étienne is the son of a laundry woman from *L'assomoir* (1877), the brother of *Nana* (1880) and of Jacques Lantier in the 1890 novel *La Bête humaine*.

49. For a longer discussion of these terms, see my essay "Ambiguity: Ambience, Ambivalence, and the Environment," *Common Knowledge* 19, no. 1 (December 2012) (Symposium: Fuzzy Studies, part 4): 88–95.

50. The relation of timber and wood to coal is complex. For now it is enough to note that when the miners want to meet without the knowledge or intervention of the managers, they do so in the forest, which they perceive to be a *zona franca*, a space that is unregulated, a wooded commons.

51. Philip Walker argues convincingly that *Germinal* reflects a "new faith" vision of a catastrophic geology that a young Zola described in an article in 1865. "He wanted to believe that more nearly perfect lands and beings were already taking shape in the deep recesses of the earth and in mankind's dreams" (Walker, 258). For Zola, the catastrophes of the mine were linked to great natural upheavals; class struggle to natural struggles.

52. The figure of the dwarf has a long and complex history in relation to mining. In some mythologies, the dwarf was assumed to have his home underground, where he guarded treasures. See my *Alchemical Mercury* for a more detailed discussion and bibliography on this topic.

53. Ansell-Pearson writes that Deleuze was interested in the relation of organismic and inorganic life, the indeterminacy of "life" itself. In a sentence that seems to hold great significance for this discussion, he notes that for Deleuze, "life is informed by the ability of its forms and expressions to hold chemical energy in a potential state and which serve as little explosives that need only a spark to set free the energy stored within them" (34).

54. See Valderrama, for example. I am grateful to Kathleen Long for this reference.

55. Peter J. Bernstein, "Ford Works on 'Sunshine Engine,'" *Chicago Daily News*, August 9, 1973, n.p.

56. www.lifegem.com. I am grateful to (and disturbed by) Vincent Bruyère for bringing Life Gem to my attention.

57. In the *Star Trek* universe, dilithium is not just two lithium atoms put together. It is an element (as shown in a periodic table: atomic weight 87 during one episode). The crystals are composed of $2(5)6$ dilithium $2(:)1$ diallosilicate $1:9:1$ heptoferranide. See Ian Steadman, "Dilithium Crystals Could Power Hypothetical, *Star Trek*-Style Warp Drive," *Wired UK*, http://www.wired.com/wiredscience/2012/10/dilithium-crystals-warp-drive/. Warp drive is a motion faster than the speed of light achieved by the warping of time and space.

58. The Order of the Golden Fleece was established in 1429 by Philippe Le Bon, Duke of Burgundy. Based on the Knights of the Round Table, the order recognized virtue and good living, faith and fraternity. The origins of the name are obscure but may be linked with the Flanders wool industry. Members wore a necklace with an emblem of the fleece. Certainly the myth of Jason and the Argonauts was invoked as part of the order's rhetoric and practices.

59. See Déprez-Masson for the illustrations to the *De re metallica*.

60. Humphry Davy (1778–1829) is important for the history of fuels. In his early career he wrote a treatise on the possible use of nitrous oxide (laughing gas) as an anesthetic for surgery. Although it didn't take off for that purpose, it became a social activity. Davy is best known for his miner safety lamp. Using **vegetable oil** as a fuel, the lamp burns a wick that is enclosed in a mesh. In this way, the lamp does not ignite **methane** (firedamp), if present. The mesh is a kind of "fuel-arrestor," like asbestos. But the lamp also had another function—to help detect the presence of methane (the flame would burn blue) or the lack of oxygen (the flame would be extinguished). The Davy lamp allowed for expanded mining in what were more dangerous areas (which, in fact, led to more accidents).

61. Jeremy Rifkin published his now-dated *Hydrogen Economy* with the subtitle *The Creation of the Worldwide Energy Web and the Redistribution of Power on Earth* in 2002.

62. Brochure, "The Florence Way," Advertising Ephemera, Box 30, "Heating and Fuel," Duke University Library Special Collections.

63. Most landfills are as much as 60 percent methane, as well as CO_2 and various contaminants.

64. *Megacities: São Paulo*, television documentary, directed by Flavia Angelico and Joseph Paterson (Washington, D.C.: National Geographic Channel, 2005), first broadcast March 28, 2006. I am grateful to Claudia Soria for bringing this documentary to my attention.

65. In 2010, the U.S. Energy Information Administration revised its protocols for estimating natural gas resources. Still, the question is murky, to say the least. In 2011, before the current boom was in full swing, an expert noted, "If the country is going to embrace natural gas as the fuel of the future, there needs to be a lot more transparency in how these estimates are calculated and a more skeptical and informed discussion about the economics of shale gas." See Urbana.

66. Neela Banerjee, "EPA Drastically Underestimates Methane Released at Drilling Sites," *Los Angeles Times*, 14 April, 2014, http://www.latimes.com/science/science now/la-sci-sn-methane-emissions-natural-gas-fracking-20140414,0,2417418.

67. According to a recent study, there may be more organic carbon available through methane hydrates than from all other existing fossil-fuel reserves combined. The biggest obstacle to its use as a fuel might be access.

68. FORD, Letter of March 26, 1916, to HF by John H. O'Neil, Dunkirk, N.Y., in Acc. 0069: Henry Ford and Son Laboratories Series, Box 1, Benzine and other chemicals (as a gasoline substitute), correspondence.

69. A company called Atlantis is placing turbines under the sea off the coast of Scotland to generate marine energy.

70. The Pennsylvania oil sites (and the use of kerosene) were primarily rural.

71. Imre Szeman comments on the relatively small genre of oil novels: "The dearth of oil in contemporary fiction is not a structuring absence that haunts the whole of

literature—an absence inescapably present through negation (standard tricks of the literary-critical trade won't save us here). . . . Instead of challenging the fiction of surplus—as well might have hoped or expected—literature participates in it just as surely as every other social narrative in the contemporary era. Ever more narrative, ever more signification, ever more grasping after social meaning; what literature shares with the Enlightenment and capitalism is the implicit longing for the plus beyond what is . . . apparent epistemic inability or unwillingness to name our energy ontologies. . . . We know where we stand with respect to energy, but we do nothing about it" (Szeman, in Yeager, 324).

72. The first filling station was constructed in Detroit in 1911. An article on the possibility of **uranium**-powered nuclear reactors for cars, written in 1954, jokes that the driver would never have an excuse to check the tire pressure or oil level. In other words, the replacement of the filling station would change the whole rhythm of driving and living. See U.F., "A quando l'energia atomica a disposizione dell'automobile?" *Motor Italia*, no. 27 (November–December 1954): 33–37.

73. The best-known version is "The Mines of Falun" by E. T. A. Hoffmann, 1819. For a wonderful analysis of different permutations of this narrative, see John Neubauer, "The Mines of Falun: Temporal Fortunes of a Romantic Myth of Time," *Studies in Romanticism* 19, no. 4 (Winter 1980): 475–95.

74. This is the kind of thinking characteristic of the neurotic driver in Italo Calvino's short story "The Petrol Pump" ("La pompa di benzina"). The narrator alternates between descriptions on the surface of modern Italy and in/through deep time: "I should have thought of it before, it's too late now. It's after twelve thirty and I didn't remember to fill up; the service stations will be closed until three. Every year two million tons of crude are brought up from the earth's crust where they have been stored for millions of centuries in the folds of rocks buried between layers of sand and clay. If I set off now there's a danger I'll run out on the way; the gauge has been warning me for quite a while that the tank is in reserve. They have been warning us for quite a while that underground global reserves can't last more than twenty years or so" (170). It's nearly impossible to drive, forward, safely, while thinking in such a manner. For post-boom Italian filling stations, see the exhibition catalogue *On the Road*, ed. Natalina Costa (Milan: 24 Ore Cultura, 2011).

75. Prentiss notes that peat was burned in blast furnaces in the British Isles in the eighteenth century and used to smelt lead ore in the Pennines. It could pack a powerful punch, in other words.

76. The narrator hides out, waiting for the Templar ritual in the Paris Conservatoire Nationale des Arts et Métiers (flanked by rue Vaucanson and rue Montgolfier!). He wonders why the little cubicle where he has chosen to hide, which he describes as "positivistic and Vernian," is placed next to a highly "emblematic [and alchemical] symbol of a lion and a serpent." In this episode we have a cluster of key elements found in *Fuel*—by chance? (Eco, 24).

77. FORD 0069, Henry Ford and Son Laboratories Series 1913–19, Box 1, Alcohol (as a gasoline substitute), correspondence.

78. During the winter of 2014, supply in the United States was unexpectedly tight due to the Polar Vortex and extremely cold temperatures. A recent brochure sent to rural areas in New York State depicts a middle-aged white couple with slightly worried expressions and a reassuring text promising that there is no shortage of propane in the natural world, only a problem of distribution.

79. Soddy (232) notes that if actinium D or thorium D had "elected" to expel an A- instead of a B-particle, the product would have been an isotope of gold instead of lead. Both thallium and **bismuth** are also close to gold.

80. FORD, Benzine and other chemicals (as a gasoline substitute), correspondence.

81. For rays as the signs of divinity, see Blaise de Vigenère: *"Car l'entendement human, selon Hermes, est comme un miroir, ou' se viennent racueiller & rabarre les clairs & lumineux rayons de la Divinite', representee a' nos sentimens par le soleil la' haute, & le feu son correspondant icy bas, lesquesls enflamment l'ame d'un ardent desire de la cognoissance & veneration de son Createur, & par consequent de l'amour d'iceluy, car l'on n'aime que ce qu'on cognoist."* ("For human understanding, according to Hermes, is like a mirror where the clear and luminous rays of Divinity are collected and refracted, represented to our senses by the sun on high and fire, its earthy counterpart; which enflame the soul with a burning desire for knowledge and veneration of our Creator and so also love for him, since one only loves what one knows" [5]). Translation mine.

82. Cited in Mark Novak, blog: www.paleofutures.com.

83. Prentiss writes: "At the deepest level, renewable energy sources transform energy from sunlight into a form that people can use because sunlight gives the water evaporation that underlies hydroelectric power, as well as the temperature differences that produce wind power and the photosynthesis that allows plants to store energy that can be released by burning plants (e.g., wood) or their by-products (e.g., ethanol)" (33).

84. The Nucleon, built as a 3/8 scale model, is on view at the Henry Ford Museum in Dearborn. It bears a clear resemblance to the Bat Car. As a brochure for the vehicle explains, it was not practicable, but "the present bulkiness and weight of nuclear reactors and attendant shielding will some day be reduced, the Nucleon is intended to probe possible design influence of atomic power in automobiles."

85. Ford imagined that the Nucleon would go for five thousand miles between charges.

86. In a letter, Bulwer-Lytton expresses some anxiety about how Vril might be misconstrued by potential readers: "I did not mean Vril for mesmerism, which I hold to be a mere branch current of the one great fluid pervading all nature. . . . Now as some bodies are charged with electricity like the torpedo or electric eel, and can never communicate that power to other bodies, so I suppose the existence of a race charged with that electricity and having acquired the art to concentrate and direct it—in a word, to be conductors of its lightnings. If you can suggest any other idea of

carrying out that idea of a destroying race, I should be glad. Probably even the notion of Vril might be more cleared from mysticism or mesmerism by being simply defined to be electricity and conducted by those staves or rods, omitting all about mesmeric passes, etc. Perhaps too, it would be safe to omit all reference to the power of communicating with the dead" (cited in Bulwer-Lytton, 167n3).

87. FORD, Experimental and Concept Vehicles Product Literature Collection, 1934–2004, Acc. 1707, Box 1.

88. Various groups are experimenting with modes of capturing wind, including through the use of long tubes rather than blades. See Bill Tucker, "Wind Power without the Mills," http://www.forbes.com/sites/billtucker/2015/05/07/wind-power-with out-the-mills/.

89. "Gelassenheit" was the German title for the 1959 essay translated into English as "Discourse on Thinking." It deals with the distinction between calculation and contemplation. The latter, which is a higher form for Heidegger, does not lead to any active or practical end. It is carried out not in any particular place (it does not require expertise), but rather on "one's home ground," which for our purposes might be akin to the hearth/fuel. Contemplation for Heidegger is an active waiting in that one must prepare for it, exercise one's capacity to do it. In *Commonwealth*, Michael Hardt and Antonio Negri critique the notion of *Gelassenheit* in its relation to work on the environment, among other areas of intervention. For them, the term means a form of power-less-ness, a withdrawal from engagement. They write: "He [Heidegger] brings phenomenology back to classical ontology not in order to develop a means to reconstruct being through human productive capacities but rather as a meditation on our telluric condition, our powerlessness, and death" (29). And for them, Heidegger influences a negative strand of biopolitics carried through into the work of Giorgio Agamben. "Agamben transposes biopolitics in a theological-political key, claiming that the only possibility of rupture with biopower resides in "inoperative" activity (*inoperosità*), a blank refusal that recalls Heidegger's notion of Gelassenheit, completely incapable of constructing an alternative" (58).

90. Ecuador is a dollarized economy and the government is currently supporting mega-mining projects.

91. As Kevin Attell notes, a great deal rests on Agamben's rereading of a phrase in Aristotle's *Metaphysics* distinguishing the relation of actuality and potentiality (*energeia* and *dynamis*). Traditional commentators of Aristotle have translated *dynamis* to mean something like "being logically possible." But Agamben insists on another meaning closer to "capability." Agamben suggests that these ancient terms are all the more significant in the present, when humanity has "grown and developed its potency [potere] to the point of imposing its power over the whole planet" (177).

BIBLIOGRAPHY

Agamben, Giorgio. *Potentialities*. Ed. and trans. Daniel Heller-Roazen. Stanford: Stanford University Press, 1999.

Agricola, Georgius. *De re metallica* (1556). Trans. Herbert Clark Hoover and Lou Henry Hoover. London: The Mining Magazine, 1912.

Ansell-Pearson, Keith. *Germinal Life: The Difference and Repetition of Deleuze*. London: Routledge, 1999.

Antonio, Robert J., ed. *Marx and Modernity: Key Readings and Commentary*. Oxford: Basil Blackwell, 2003.

Antonioni, Michelangelo, dir. *Red Desert* (1964).

Apollonius of Rhodes. *Argonautika* (3rd century BCE).

Apollonius Rhodius. *The Argonautica*. Ed. and trans. R. C. Seaton. Cambridge, Mass.: Harvard University Press, 1912.

Aristotle. *De anima*. Trans. J. A. Smith. http://classics.mit.edu/Aristotle/soul.html.

Attell, Kevin. "Potentiality, Actuality, Constituent Power." *Diacritics [Contemporary Italian Thought 1]* 39, no. 3 (Fall 2009): 35–54.

———. *Giorgio Agamben: Beyond the Threshold of Deconstruction*. New York: Fordham University Press, 2014.

Bacigalupi, Paolo. *The Windup Girl*. San Francisco: Night Shade Books, 2011.

Barthes, Roland. *Mythologies*. Trans. Annette Lavers. New York: Hill and Wang, 1972.

Benjamin, Walter. *Illuminations*. Trans. Harry Zohn. London: Fontana, 1992.

Bennett, Jane. *Vibrant Matters: A Political Ecology of Things*. Durham: Duke University Press, 2010.

Bensaïd, Daniel. *Les dépossédés: Karl Marx, les voleurs de bois et le droit des pauvres*. Paris: La fabrique, 2007.

Bogost, Ian. *Alien Phenomenology, or What It's Like to Be a Thing*. Minneapolis: University of Minnesota Press, 2012.

Bulwer-Lytton, Edward. *The Coming Race*. Middletown, Conn.: Wesleyan University Press, 2005.

Calvino, Italo. "The Petrol Pump." In *Numbers in the Dark and Other Stories*. Trans. Tim Parks. New York: Pantheon Books, 1995.

Campos, Luis. "The Birth of Living Radium." *Representations* 97 (Winter 2007): 1–27.

The Chemical Wedding of Christian Rosenkreutz. Trans. Joscelyn Godwin. Introduction and commentary by Adam McLean. Newburyport, Mass., Phanes Press, 1991.

Chesneaux, Jean. *The Political and Social Ideas of Jules Verne*. Trans. Thomas Wikeley. London: Thames and Hudson, 1972.

Clark, B., and J. B. Foster. "William Stanley Jevons and the Coal Question: An Introduction to Jevons's 'Of the Economy of Fuel.'" *Organization & Environment* 14 (2001): 93–98.

Clarke, Bruce. *Energy Forms: Allegory and Science in the Era of Classical Thermodynamics*. Ann Arbor: University of Michigan Press, 2001.

Cohen, Tom, ed. *Telemorphosis: Theory in the Era of Climate Change*, vol. 1. Ann Arbor: Open University Press, 2012.

Coopersmith, Jennifer. *Energy: The Subtle Concept*. Oxford Scholarship Online, 2010.

Crease, Robert. "Energy in the History and Philosophy of Science." *Elsevier Encyclopedia of Energy*. 2004. Web.

Daston, Lorraine, and Peter Galison. *Objectivity*. Cambridge, Mass.: Zone Books, 2010.

Déprez-Masson, Marie-Claude. *Technique, mot et image: Le De re metallica d'Agricola*. Brepols: Turnhout, 2006.

Denatured Alcohol: Hearings before the Committee on Agriculture, House of Representatives, 64th Cong., 1st sess., on H.R. 11256. July 14, 1916.

Derrida, Jacques. *Given Time: I. Counterfeit Money*. Trans. Peggy Kamuf. Chicago: University of Chicago Press, 1992.

———. "The Deconstruction of Actuality." In *Negotiations: Interventions and Interviews, 1971–2001*. Ed. and trans. Elizabeth Rottenberg. Stanford: Stanford University Press, 2001.

Dickens, Charles. *Bleak House*. New York: Barnes and Noble, 2005.

Doray, Bernard. *From Taylorism to Fordism: A Rational Madness*. London: Free Association Books, 1988.

Eco, Umberto. *Il pendolo di Foucault*. Milan: Bompiani, 1998.

Eliot, George. *Silas Marner* (1861).

Elsevier Encyclopedia of Energy. Editor in chief Cutler J. Cleveland. Elsevier, 2004. Web.

Evans, Arthur. *Jules Verne Rediscovered: Didacticism and the Scientific Novel*. Westport, Conn.: Greenwood Press, 1988.

Evelyn, John. *Sylva, or a Discourse of Forest-Trees and the Propagation of Timber in his Majesty's Dominions* (1664). London: Doubleday, 1908.

Faivre, Antoine. *The Golden Fleece and Alchemy*. Albany: State University of New York Press, 1993.

Flamel, Nicolas. *His Exposition of the Hieroglyphical Figures which he caused to be painted upon an Arch in St. Innocents Church-Yard, in Paris.* Trans. Eirenaeus Orandus. London, 1624.

Fletcher, Angus. *Allegory: Theory of a Symbolic Mode.* Ithaca: Cornell University Press, 1964.

Floudas, Dimitrios, et al. "The Paleozoic Origin of Enzymatic Lignin Decomposition Reconstructed from 31 Fungal Genomes." *Science* 29 (June 2012): 1715–19. DOI:10.1126.

Gilbert, William. *De magnete* (1600).

———. *On the Loadstone and Magnetic Bodies and on the Great Magnet The Earth* (1600). Trans. P. Fluery Mottelay. New York: John Wiley and Sons, 1893.

Ghosh, Amitav. "Petrofiction: The Oil Encounter and the Novel" (1992), reprinted in his *Incendiary Circumstances: A Chronicle of the Turmoil of Our Times.* Boston: Houghton Mifflin, 2005.

Grandin, Greg. *Fordlandia: The Rise and Fall of Henry Ford's Forgotten Jungle City.* New York: Metropolitan Books, 2009.

Greer, Diane. "Creating Cellulosic Ethanol: Spinning Straw into Fuel." *BioCycle* (April 2005).

Hack, John. *Prehistoric Coal Mining in the Jeddito Valley, Arizona.* Cambridge, Mass.: Peabody Museum, 1942.

Hardt, Michael, and Antonio Negri. *Commonwealth.* Cambridge, Mass.: Belknap Press of Harvard University Press, 2011.

Heidegger, Martin. *Discourse on Thinking: A Translation of "Gelassenheit" by John M. Anderson and E. Hans Freund* (1959). New York: Harper and Row, 1966.

———. "The Question Concerning Technology." In *The Question Concerning Technology and Other Essays.* Trans. William Lovitt. New York: Garland, 1977.

Homer. *Odyssey.* Trans. Richard Lattimore. New York: Harper and Row, 1967.

Jones, Duncan, dir. *Moon* (2008).

Kosinski, Joseph, dir. *Oblivion* (2013).

Krauss, Lawrence. *The Physics of Star Trek.* New York: Basic Books, 1995.

LeMenager, Stephanie. *Living Oil: Petroleum Culture in the American Century.* Oxford: Oxford University Press, 2014.

Lyotard, Jean-François. "Oikos" (1988). In *Political Writings.* Trans. Bill Readings with Kevin Paul Geiman. Minneapolis: University of Minnesota Press, 1993.

———. *Discourse, Figure.* Trans. Antony Hudek and Mary Lydon. Minneapolis: University of Minnesota Press, 2011.

Macherey, Pierre. *A Theory of Literary Production* (1966). Trans. Geoffrey Wall. London: Routledge and Kegan Paul, 1978.

Marinetti, Filippo Tommaso. "La nuova religione-morale della velocità" (1916). In *Teoria e invenzione futurista.* Ed. Luciano De Maria. Milan: Arnoldo Mondadori, 1968.

Marx, Karl. *Capital*. Vol I. Trans. Samuel Moore and Edward Aveling. Ed. Frederick Engels. Moscow: Progress Publishers, 1967.

———. *Economic and Philosophic Manuscripts of 1844*. Trans. Martin Milligan. Moscow: Progress Publishers, 1956.

———. *Selected Writings*. Ed. David McLellan. Oxford: Oxford University Press, 1977.

McPhee, John. *Annals of the Former World*. New York: Farrar, Straus and Giroux, 2000.

Mitchell, Timothy. *Carbon Democracy: Political Power in the Age of Oil*. New York: Verso Books, 2011.

Miyazaki, Hirokazu. *The Method of Hope: Anthropology, Philosophy, and Fijian Knowledge*. Stanford: Stanford University Press, 2004.

Negri, Antonio. *Insurgencies*. Trans. Maurizia Boscagli. Minneapolis: University of Minnesota Press, 1999.

Nikiforuk, Andrew. *Tar Sands: Dirty Oil and the Future of a Continent*. Vancouver: Greystone Books, 2010.

———. *The Energy of Slaves: Oil and the New Servitude*. Vancouver: Greystone Books, 2014.

Pasolini, Pier Paolo. *Petrolio*. Turin: Einaudi, 1992.

Pinkus, Karen. *Alchemical Mercury: A Theory of Ambivalence*. Stanford: Stanford University Press, 2009.

Prentiss, Mara. *Energy Revolution: The Physics and the Promise of Efficient Technology*. Cambridge, Mass.: Belknap Press of Harvard University Press, 2015.

Price, Derek de Solla. "Automata and the Origins of Mechanism and Mechanistic Philosophy." *Technology and Culture* 5, no. 1 (Winter 1964): 12–23.

Raffles, Hugh. *Insectopedia*. New York: Pantheon Books, 2010.

Rifkin, Jeremy. *The Hydrogen Economy*. New York: Jeremy P. Tarcher/Putnam, 2002.

Robinson, Kim Stanley. *Forty Signs of Rain*. New York: Bantam Books, 2004.

Rutherford, Ernest. *The Newer Alchemy. Based on the Hendry Sidgwick Memorial Lecture Delivered at Newnham College Cambridge, November, 1936*. New York: Macmillan, 1937.

Sieferle, Rolf Peter. *The Subterranean Forest: Energy Systems and the Industrial Revolution* (1982). Trans. Michael P. Osman. Cambridge: White Horse Press, 2001.

Sigfusson, Thorsteinn. *Planet Hydrogen: The Taming of the Proton*. Oxford: Coxmoor, 2008.

Sinclair, Upton. *King Coal*. New York: Bantam Classics, 1994.

———. *Oil!* New York: Penguin Books, 2007.

Smith, Crosbie. *The Science of Energy: A Cultural History of Energy Physics in Victorian Britain*. Chicago: University of Chicago Press, 1998.

Snyder, James G. "The Theory of *materia prima* in Marsilio Ficino's Platonic Theology." *Vivarium* 46, no. 2 (2008): 192–221.

Soddy, Frederick. *The Interpretation of Radium and the Structure of the Atom* (1908), 4th edition. New York: Putnam's, 1920.

Spitzer, Leo. "Milieu and Ambience." In *Essays in Historical Semantics*. New York: Russell & Russell, 1968.

Stimilli, Davide. *The Face of Immortality: Physiognomy and Criticism*. Albany: State University of New York Press, 2005.

Tiffany, Daniel. *Toy Medium: Materialism and Modern Lyric*. Berkeley: University of California Press, 2000.

Unwin, Timothy. *Jules Verne: Journeys in Writing*. Liverpool: Liverpool University Press, 2006.

Urbana, Ian. "Geologists Sharply Cut Estimate of Shale Gas." *New York Times,* August 25, 2011.

Valderrama, Ignacio Miguel Pascual, and Jacquín Pérez-Pariente. "Alchemy at the Service of Mining Technology in Seventeenth-Century Europe, According to the Works of Martine de Bertereau and Jean du Castelet." *Bulletin of the History of Chemistry* 37, no. 1 (2012): 1–13.

Vernant, Jean-Pierre, "Hestia-Hermès: Sur l'expression religieuse de l'espace et du mouvement chez les Grecs." *L'homme* 3, no. 3 (September–December 1963): 12–50.

Verne, Jules. *Twenty Thousand Leagues under the Sea* (1870). Trans. Walter James Miller and Frederick Paul Walter. Annapolis: Naval Institute Press, 1993.

———. *The Mysterious Island* (1874). Trans. Jordan Stump. New York: Random House, 2001.

———. *Robur the Conquerer* (1886). *Works of Jules Verne*, vol. 14. Ed. Charles Horne. New York: Vincent Parke and Co., 1911.

Vigenère, Blaise de. *Traité du feu et du sel* (1600). Paris: Jobert, 1976.

Walker, Philip. "*Germinal* and Zola's Youthful 'New Faith' Based on Geology." *Symposium* 36, no. 3 (1982): 257–72.

Wells, H. G. *The World Set Free* (1914). London: Hogarth Press, 1988.

Wik, Reynold. *Henry Ford and Grassroots America*. Ann Arbor: University of Michigan Press, 1972.

Yeager, Patricia. "Editor's Column: Literature in the Ages of Wood, Tallow, Coal, Whale Oil, Gasoline, Atomic Power, and Other Energy Sources." Contributions by Ken Hiltner, Saree Makdisi, Vin Nardizzi, Laurie Shannon, Imre Szeman, and Michael Ziser. *PMLA* 126, no. 2 (March 2011): 305–26.

Zola, Émile. *Germinal*. Trans. Peter Collier. Oxford: Oxford University Press, 2008.

KAREN PINKUS is professor of Italian and comparative literature at Cornell University and a member of the Faculty Advisory Board of the Atkinson Center for a Sustainable Future. She teaches and writes on climate change and the humanities. She is author of *Bodily Regimes: Italian Advertising under Fascism* (Minnesota, 1995); *Picturing Silence: Emblem, Language, Counter-Reformation Materiality*; *The Montesi Scandal: The Death of Wilma Montesi and the Birth of the Paparazzi in Fellini's Rome*; and *Alchemical Mercury: A Theory of Ambivalence*.

(continued from page ii)